WITHDRAWN

Exploration of Africa: The Emerging Nations

NORTH AFRICA

Exploration of Africa: The Emerging Nations

North Africa

John G. Hall

Introductory Essay by
Dr. Richard E. Leakey
Chairman, Wildlife Clubs
of Kenya Association

✣

Afterword by
Deirdre Shields

CHELSEA HOUSE PUBLISHERS
Philadelphia
In association with Covos Day Books, South Africa

CHELSEA HOUSE PUBLISHERS

EDITOR IN CHIEF Sally Cheney
DIRECTOR OF PRODUCTION Kim Shinners
CREATIVE MANAGER Takeshi Takahashi
MANUFACTURING MANAGER Diann Grasse

Staff for NORTH AFRICA

EDITOR Lee M. Marcott
PRODUCTION ASSISTANT Jaimie Winkler
COVER DESIGN Emiliano Begnardi
SERIES DESIGNER Keith Trego

© 2003 by Chelsea House Publishers, a subsidiary of Haights Cross Communications. All rights reserved. Printed and bound in the United States of America.

The Chelsea House World Wide Web address is http://www.chelseahouse.com

First Printing

1 3 5 7 9 8 6 4 2

All photographs in this book © Royal Geographical Society. No reproduction of images without permission.

Royal Geographical Society
1 Kensington Gore
London SW7 2AR

Library of Congress Cataloging-in-Publication Data

Hall, John G., 1950-
  North Africa / John G. Hall ; introductory essay by Richard Leakey; afterword by Deidre Shields.
    v. cm. — (Exploration of Africa, the emerging nations) Summary: A history of the land and people of the geographic entity known as North Africa.
    Includes bibliographical references and index.
    Contents: Algeria—Morocco—Tunisia—Libya.
  ISBN 0-7910-5746-1
  1. Africa, North—Juvenile literature. [1. Africa, North.] I. Title.
  II. Series.
  DT185 .H35 2002
  961—dc21
                                                    2002008251

The photographs in this book are from the Royal Geographical Society Picture Library. Most are being published for the first time.

The Royal Geographical Society Picture Library provides an unrivaled source of over half a million images of peoples and landscapes from around the globe. Photographs date from the 1840s onwards on a variety of subjects including the British Colonial Empire, deserts, exploration, indigenous peoples, landscapes, remote destinations, and travel.

Photography, beginning with the daguerreotype in 1839, is only marginally younger than the Society, which encouraged its explorers to use the new medium from its earliest days. From the remarkable mid-19th century black-and-white photographs to color transparencies of the late 20th century, the focus of the collection is not the generic stock shot but the portrayal of man's resilience, adaptability, and mobility in remote parts of the world.

In organizing this project, we have incurred many debts of gratitude. Our first, though, is to the professional staff of the Picture Library for their generous assistance, especially to Joanna Scadden, Picture Library Manager.

CONTENTS

	THE DARK CONTINENT *Dr. Richard E. Leakey*	7
	INTRODUCTION	15
1	ALGERIA	21
2	MOROCCO	39
3	TUNISIA	67
4	LIBYA	87
5	CONCLUSION	109
	WORLD WITHOUT END *Deirdre Shields*	112
	CHRONOLOGY	117
	FURTHER READING	121
	RESOURCES USDED	122
	INDEX	123

Exploration of Africa: The Emerging Nations

Angola
Central and East Africa
The Congo
Egypt
Ethiopia
Nigeria
North Africa
South Africa
Southeast Africa
Sudan
West Africa

THE DARK CONTINENT

DR. RICHARD E. LEAKEY

THE CONCEPT OF AFRICAN exploration has been greatly influenced by the hero status given to the European adventurers and missionaries who went off to Africa in the last century. Their travels and travails were certainly extraordinary and nobody can help but be impressed by the tremendous physical and intellectual courage that was so much a characteristic of people such as Livingstone, Stanley, Speke, and Baker, to name just a few. The challenges and rewards that Africa offered, both in terms of commerce and also "saved souls," inspired people to take incredible risks and endure personal suffering to a degree that was probably unique to the exploration of Africa.

I myself was fortunate enough to have had the opportunity to organize one or two minor expeditions to remote spots in Africa where there were no roads or airfields and marching with porters and/or camels was the best option at the time. I have also had the thrill of being with people untouched and often unmoved by contact with Western or other technologically based cultures, and these experiences remain for me amongst the most exciting and salutary of my life. With the contemporary revolution in technology, there will be few if any such opportunities again. Indeed I often find myself slightly saddened by the realization that were life ever discovered on another planet, exploration would doubtless be done by remote sensing and making full use of artificial, digital intelligence. At least it is unlikely to be in my lifetime and this is a relief!

NORTH AFRICA

Notwithstanding all of this, I believe that the age of exploration and discovery in Africa is far from over. The future offers incredible opportunities for new discoveries that will push back the frontiers of knowledge. This endeavor will of course not involve exotic and arduous journeys into malaria-infested tropical swamps, but it will certainly require dedication, team work, public support, and a conviction that the rewards to be gained will more than justify the efforts and investment.

Early Explorers

Many of us were raised and educated at school with the belief that Africa, the so-called Dark Continent, was actually discovered by early European travelers and explorers. The date of this "discovery" is difficult to establish, and anyway a distinction has always had to be drawn between northern Africa and the vast area south of the Sahara. The Romans certainly had information about the continent's interior as did others such as the Greeks. A diverse range of traders ventured down both the west coast and the east coast from at least the ninth century, and by the tenth century Islam had taken root in a number of new towns and settlements established by Persian and Arab interests along the eastern tropical shores. Trans-African trade was probably under way well before this time, perhaps partly stimulated by external interests.

Close to the beginning of the first millennium, early Christians were establishing the Coptic church in the ancient kingdom of Ethiopia and at other coastal settlements along Africa's northern Mediterranean coast. Along the west coast of Africa, European trade in gold, ivory, and people was well established by the sixteenth century. Several hundred years later, early in the 19th century, the systematic penetration and geographical exploration of Africa was undertaken by Europeans seeking geographical knowledge and territory and looking for opportunities not only for commerce but for the chance to spread the Gospel. The extraordinary narratives of some of the journeys of early European travelers and adventurers in Africa are a vivid reminder of just how recently Africa has become embroiled in the power struggles and vested interests of non-Africans.

THE DARK CONTINENT

Africa's Gift to the World

My own preoccupation over the past thirty years has been to study human prehistory, and from this perspective it is very clear that Africa was never "discovered" in the sense in which so many people have been and, perhaps, still are being taught. Rather, it was Africans themselves who found that there was a world beyond their shores.

Prior to about two million years ago, the only humans or protohumans in existence were confined to Africa; as yet, the remaining world had not been exposed to this strange mammalian species, which in time came to dominate the entire planet. It is no trivial matter to recognize the cultural implications that arise from this entirely different perspective of Africa and its relationship to the rest of humanity.

How many of the world's population grow up knowing that it was in fact African people who first moved and settled in southern Europe and Central Asia and migrated to the Far East? How many know that Africa's principal contribution to the world is in fact humanity itself? These concepts are quite different from the notion that Africa was only "discovered" in the past few hundred years and will surely change the commonly held idea that somehow Africa is a "laggard," late to come onto the world stage.

It could be argued that our early human forebears—the *Homo erectus* who moved out of Africa—have little or no bearing on the contemporary world and its problems. I disagree and believe that the often pejorative thoughts that are associated with the Dark Continent and dark skins, as well as with the general sense that Africans are somehow outside the mainstream of human achievement, would be entirely negated by the full acceptance of a universal African heritage for all of humanity. This, after all, is the truth that has now been firmly established by scientific inquiry.

The study of human origins and prehistory will surely continue to be important in a number of regions of Africa and this research must continue to rank high on the list of relevant ongoing exploration and discovery. There is still much to be learned about the early stages of human development, and the age of the "first humans"—the first bipedal apes—has not been firmly established. The current hypothesis is that prior to five million years ago there were no bipeds, and this

North Africa

would mean that humankind is only five million years old. Beyond Africa, there were no humans until just two million years ago, and this is a consideration that political leaders and people as a whole need to bear in mind.

Recent History

When it comes to the relatively recent history of Africa's contemporary people, there is still considerable ignorance. The evidence suggests that there were major migrations of people within the continent during the past 5,000 years, and the impact of the introduction of domestic stock must have been quite considerable on the way of life of many of Africa's people. Early settlements and the beginnings of nation states are, as yet, poorly researched and recorded. Although archaeological studies have been undertaken in Africa for well over a hundred years, there remain more questions than answers.

One question of universal interest concerns the origin and inspiration for the civilization of early Egypt. The Nile has, of course, offered opportunities for contacts between the heart of Africa and the Mediterranean seacoast, but very little is known about human settlement and civilization in the upper reaches of the Blue and White Nile between 4,000 and 10,000 years ago. We do know that the present Sahara Desert is only about 10,000 years old; before this Central Africa was wetter and more fertile, and research findings have shown that it was only during the past 10,000 years that Lake Turkana in the northern Kenya was isolated from the Nile system. When connected, it would have been an excellent connection between the heartland of the continent and the Mediterranean.

Another question focuses on the extensive stone-walled villages and towns in Southern Africa. The Great Zimbabwe is but one of thousands of standing monuments in East, Central, and Southern Africa that attest to considerable human endeavor in Africa long before contact with Europe or Arabia. The Neolithic period and Iron Age still offer very great opportunities for exploration and discovery.

As an example of the importance of history, let us look at the modern South Africa where a visitor might still be struck by the not-too-subtle representation of a past that, until a few years ago, only "began" with the arrival of Dutch settlers some 400 years back. There are, of

THE DARK CONTINENT

course, many pre-Dutch sites, including extensive fortified towns where kingdoms and nation states had thrived hundreds of years before contact with Europe; but this evidence has been poorly documented and even more poorly portrayed.

Few need to be reminded of the sparseness of Africa's precolonial written history. There are countless cultures and historical narratives that have been recorded only as oral history and legend. As postcolonial Africa further consolidates itself, history must be reviewed and deepened to incorporate the realities of precolonial human settlement as well as foreign contact. Africa's identity and self-respect is closely linked to this.

One of the great tragedies is that African history was of little interest to the early European travelers who were in a hurry and had no brief to document the details of the people they came across during their travels. In the basements of countless European museums, there are stacked shelves of African "curios"—objects taken from the people but seldom documented in terms of the objects' use, customs, and history.

There is surely an opportunity here for contemporary scholars to do something. While much of Africa's precolonial past has been obscured by the slave trade, colonialism, evangelism, and modernization, there remains an opportunity, at least in some parts of the continent, to record what still exists. This has to be one of the most vital frontiers for African exploration and discovery as we approach the end of this millennium. Some of the work will require trips to the field, but great gains could be achieved by a systematic and coordinated effort to record the inventories of European museums and archives. The Royal Geographical Society could well play a leading role in this chapter of African exploration. The compilation of a central data bank on what is known and what exists would, if based on a coordinated initiative to record the customs and social organization of Africa's remaining indigenous peoples, be a huge contribution to the heritage of humankind.

Medicines and Foods

On the African continent itself, there remain countless other areas for exploration and discovery. Such endeavors will be achieved without the fanfare of great expeditions and high adventure as was the case during the last century and they should, as far as possible, involve exploration and discovery of African frontiers by Africans themselves. These frontiers are not geographic: they are boundaries of knowledge in the sphere of Africa's home-grown cultures and natural world.

Indigenous knowledge is a very poorly documented subject in many parts of the world, and Africa is a prime example of a continent where centuries of accumulated local knowledge is rapidly disappearing in the face of modernization. I believe, for example, that there is much to be learned about the use of wild African plants for both medicinal and nutritional purposes. Such knowledge, kept to a large extent as the experience and memory of elders in various indigenous communities, could potentially have far-reaching benefits for Africa and for humanity as a whole.

The importance of new remedies based on age-old medicines cannot be underestimated. Over the past two decades, international companies have begun to take note and to exploit certain African plants for pharmacological preparations. All too often, Africa has not been the beneficiary of these "discoveries," which are, in most instances, nothing more than the refinement and improvement of traditional African medicine. The opportunities for exploration and discovery in this area are immense and will have assured economic return on investment. One can only hope that such work will be in partnership with the people of Africa and not at the expense of the continent's best interests.

Within the same context, there is much to be learned about the traditional knowledge of the thousands of plants that have been utilized by different African communities for food. The contemporary world has become almost entirely dependent, in terms of staple foods, on the cultivation of only six principal plants: corn, wheat, rice, yams, potatoes, and bananas. This cannot be a secure basis to guarantee the food requirements of more than five billion people.

Many traditional food plants in Africa are drought resistant and might well offer new alternatives for large-scale agricultural development in the years to come. Crucial to this development is finding out what African people used before exotics were introduced. In some rural areas of the continent, it is still possible to learn about much of this by talking to the older generation. It is certainly a great shame that some of the early European travelers in Africa were ill equipped to study and record details of diet and traditional plant use, but I am sure that, although it is late, it is not too late. The compilation of a pan-African database on what is known about the use of the continent's plant resources is a vital matter requiring action.

Vanishing Species

In the same spirit, there is as yet a very incomplete inventory of the continent's other species. The inevitable trend of bringing land into productive management is resulting in the loss of unknown but undoubtedly large numbers of species. This genetic resource may be invaluable to the future of Africa and indeed humankind, and there really is a need for coordinated efforts to record and understand the continent's biodiversity.

In recent years important advances have been made in the study of tropical ecosystems in Central and South America, and I am sure that similar endeavors in Africa would be rewarding. At present, Africa's semi-arid and highland ecosystems are better understood than the more diverse and complex lowland forests, which are themselves under particular threat from loggers and farmers. The challenges of exploring the biodiversity of the upper canopy in the tropical forests, using the same techniques that are now used in Central American forests, are fantastic and might also lead to eco-tourist developments for these areas in the future.

It is indeed an irony that huge amounts of money are being spent by the advanced nations in an effort to discover life beyond our own planet, while at the same time nobody on this planet knows the extent and variety of life here at home. The tropics are especially relevant in this regard and one can only hope that Africa will become the focus of renewed efforts of research on biodiversity and tropical ecology.

An Afrocentric View

Overall, the history of Africa has been presented from an entirely Eurocentric or even Caucasocentric perspective, and until recently this has not been adequately reviewed. The penetration of Africa, especially during the last century, was important in its own way; but today the realities of African history, art, culture, and politics are better known. The time has come to regard African history in terms of what has happened in Africa itself, rather than simply in terms of what non-African individuals did when they first traveled to the continent.

Berber Girls, c. 1890 *Berbers are descended from the indigenous inhabitants of North Africa. Today, there are about 15 million Berbers. They speak a variety of Berber dialects. Many follow Arabic customs and traditions.*

Most Berbers depend on herding and farming for a living. They usually live in compact villages in rugged, mountain areas. Some Berber-speaking groups, such as the Tuareg of the Sahara, roam the desert with herds of camels, cows, goats, and sheep.

Arab invasions of North Africa began in the 600s. Under Arab influence, most Berbers converted to Islam. During the 700s, Muslim Berbers joined with Arabs in conquering Spain. By the mid eleventh-century, desert Berbers had organized the Almoravid Empire. At its height, this empire ruled the regions that are now Morocco, western Algeria, and southern Spain. In about 1150, mountain Berbers overthrew the Almoravids and established the Almohad Empire. The Almohads eventually controlled all of what are now Algeria, Morocco, Tunisia, and part of Spain. This large domain split apart during the first half of the thirteenth-century.

Meanwhile, Arabs continued to move into North Africa. In time they occupied most of the coastal region. Berbers there became absorbed into Arab society, and today Berber languages, traditions, and customs remain only in the mountains and deserts, areas that were relatively isolated from the Arabs.

Introduction

North Africa is an elusive term. As a geographical entity, the region has no single accepted definition. Some definitions are imposed by outsiders, as during North Africa's long colonial history; others originate from the people who inhabit the land. Therefore, any discussion of the region has to take into consideration the point of view of the discussants.

For example, seen from across the Mediterranean Sea in Europe, from which it is separated by eight miles at the Strait of Gibraltar and eighty-five miles at the Strait of Sicily, North Africa has often been called the *Barbary States* or simply *Barbary,* after the name of the indigenous Berbers, who have lived in North Africa since earliest recorded history. References to Berbers date from about 3000 B.C. and occur frequently in ancient Egyptian, Greek, and Roman writings. For many centuries the Berbers inhabited the coast of North Africa from Egypt to the Atlantic Ocean. The term *Berber* is probably derived from the Roman term for "barbarians," *barbara*. Some have claimed the word also means "outcast," or those from the land of Ber, who was the son of Ham, who in turn was the son of Noah. More important than the meaning of *Berber,* though, is the meaning of the name the Berbers use when referring to themselves. That name is *Imazighan,* which can be translated as "free men" or "free people"—certainly an accurate description of these early inhabitants, who throughout the successive invasions by Carthaginians, Romans, Vandals, Arabs, and French have maintained a remarkable degree of autonomy and independence, characteristics that continue to define them. Today the largest populations of the estimated 15 million people of Berber heritage are found in Algeria and Morocco, but significant populations also live in Tunisia and Libya.

Among the imposing landscapes of North Africa are the *Atlas Lands* of Algeria, Morocco, and Tunisia, so named for the long mountain ranges, the Atlas Mountains, that dominate their northern sides. Although all three countries, especially Algeria, are bordered by a sizable portion of the Sahara, the largest desert in the world, it is the Atlas Mountains that remain the point of reference for most outsiders. It is a different story farther east, in Libya. Here the desert becomes a recurring motif. In fact, only the northwestern and northeastern parts of the country, known as Tripolitania and Cyrenaica, respectively, are outside the desert.

The view of North Africa from across the Mediterranean Sea in Europe is drastically different than that seen through Arab eyes. Arab peoples call North Africa the *Maghrib,* literally translated as "the land and place of the sunset" or "the land of the setting sun" or simply "the West." For the early Arabic invaders this region was the "island of the West," or *Jazirat al Maghrib,* "the land between the sea and sand," referring to the Sahara and the Mediterranean Sea. The extensive area of the Maghrib is frequently differentiated into the *Maghrib al-Adna,* or the "Near West," comprising Libya and Tunisia, the *Maghrib al-Ausat,* or the "Middle West," which includes Algeria, and the *Maghrib al-Aqsa,* or the "Far West," which includes Morocco. The Atlas Lands are often graphically called *Djezirah (Jazirat) al-Maghrib,* or the "Western Isle," signifying their isolated position above the empty Sahara and the Mediterranean Sea.

Although the exact geographical meaning of the term *North Africa* differs according to various views, for the purpose of this book it is considered to coincide with the area collectively known to the Arab and French-speaking world as the Maghrib, comprising present-day Algeria, Morocco, Tunisia, and Libya.

Introduction

Moorish Cafe, Sahara Desert, 1896

Despite some lingering debate, there seems to be a general consensus today that the Maghrib is not an artificial construct, since the region has thousands of years of culture, history, religion, and language in common.

The experience of a shared geography should be added to the foregoing list because, aside from religion, geography has exerted the greatest influence on the lives of the people who live in North Africa. The region has a unique transitional position—between the Mediterranean and the Sahara, between Europe and Africa, and between the west and east. As a result of its peculiar geographical circumstances, and its resultant isolation, North Africa is, in fact, neither entirely African nor

Nomad Encampment c. 1880 *Although it is as large as the United States, the Sahara contains less than one person per square mile. Huge areas are wholly empty. In the desert proper, life is confined to the oases, where it is possible to cultivate the date palm and some fruit trees. Wherever meager vegetation can support grazing animals or where reliable water sources occur, scattered clusters of people have survived in a fragile balance with one of the harshest environments in the world.*

Archaeological evidence suggests that the Sahara was inhabited by diverse peoples about 6,000 years ago. However, it is thought that many camel-mounted nomads entered the desert to avoid the anarchy and warfare of the late Roman period in North Africa.

European nor Middle Eastern, but rather has its own distinctive identity based on an internal diversity that incorporates elements from all of these cultures. One writer has described it as a mosaic that at a distance produces a harmonious, unified image, but upon closer inspection reveals itself to be a mass of varied adamantine stones. Hundreds of potentially, if not deter-

Introduction

minedly, independent regions make up this virtual island surrounded by the three seas of the Mediterranean, the Atlantic, and to the south, the sea of sand called the Sahara.

North Africa is a land of paradox, with its coexistent uniformity and diversity. It is also a land of ancient kingdoms, myths, and legends, rich in tradition, with a long and varied history peopled by epic heroes, fabled queens, and immortal warriors. Eight centuries before the birth of Christ, the countries of North Africa—Algeria, Morocco, Tunisia, and Libya—became the stage for one of the most remarkable stories ever told and some of the most memorable characters that have ever unveiled the passions of their lives.

Street Scene, Algiers, c. 1923 *A group of women walking down a street in Algiers. Their puffy pantaloons and covered faces evoke the age of the harem and the* Arabian Nights. *However, in terms of the development of Islam and its influence, the segregated harem and the subordinate status of women was a relatively late development. The teachings of Muhammad gave women considerable freedom, and the Koran requires only modesty in dress.*

1

ALGERIA

Algeria is the largest of the North African countries, covering an area of approximately 1 million square miles, or 2,381,741 square kilometers. This is three times the size of the state of Texas and four times the size of France. Only about one-eighth of this vast country is fertile enough to sustain life, however. The Sahara Desert has claimed the rest.

Algeria is officially called the *Democratic and Popular Republic of Algeria.* It is a republic of western North Africa, bordered on the north by the Mediterranean Sea; on the east by Tunisia and Libya; on the south by Niger, Mali, and Mauritania; and on the west by Morocco.

One of the recurring themes of Algerian history has been the people's struggle to create a sense of national unity based on the shared and unique characteristics of the country's history, and to win the right to self-determination for their nation and "all of the people" within the diverse ethnic groups who reside within its borders. Even a brief examination of its history will reveal that this has been a long and costly struggle.

During the early morning hours of All Saints' Day, November 1, 1954, members of the *Front de Liberation National,* or *National Liberation Front,* launched carefully coordinated bombing attacks throughout Algeria. The military offensive was carried out by

North Africa

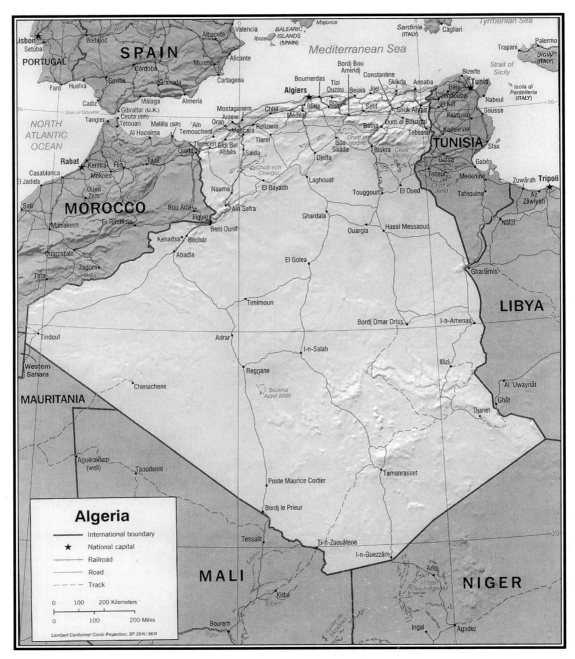

Modern map of Algeria

Algeria

Biskra, c. 1880 *Biskra is a group of oases in northern Algeria on the northern edge of Sahara. The surrounding area is extremely dry. Dates are the principal crop of the region. Today, Biskra has become a winter resort town of broad, tree-lined streets, hotels, shops, and public gardens. It is the site of a well-known modern health spa.*

Maquisards, or guerrilla bands who targeted military installations, police posts, warehouses, communications facilities, and public utilities. This insurrection was meant as a slap across the face of French Colonial rule in North Africa, especially in Algeria. It had been a long time in coming. French troops had occupied Algiers, Algeria's oldest, largest, and most historic city in 1830, after more than thirty years of increasing animosity between the *deys* (rulers) of Algeria and a succession of French governments. That occupation inaugurated 132 years of French colonial rule in Algeria.

If the chain reaction of explosions on that November day in 1954 were not enough to drive their point home, members of

Nomad Encampment at the Oasis of Biskra, c. 1880

the National Liberation Front finished the job by making a broadcast from Cairo, Egypt, its base of operations, in which they called on all Muslims in Algeria to join in a national struggle for the "restoration of the Algerian state, sovereign, democratic, and social, within the framework of the principles of Islam." It was a bold move, unprecedented in its courageousness and far-reaching in its impact. It caught the French government, and the rest of the world, completely by surprise.

The French minister of the interior, socialist François Mitterrand, responded with anger and more than just a hint of bitterness: "The only possible negotiation is war." Although his declaration left no doubt regarding the French government's attitude, it was

Algeria

the reaction of Premier Pierre Mendes-France that set the tone of French policy for the next five years. On November 12, 1954, he addressed the French National Assembly with a well-prepared speech designed to alleviate the country's growing concerns over the impending crisis in Algeria. His remarks came at a crucial time in French colonial history. Following World War II, from 1946 to 1954, the French government had engaged in an armed conflict against the *Democratic Republic of Vietnam* (DRV), known as the *First Indochina War*. Indochina, in this context, refers to Cambodia, Laos, and Vietnam, which were united under French rule as the *Indochinese Union,* commonly known as *French Indochina.* France created this union from its territories in Cambodia and Vietnam in 1887, incorporating Laos in 1893.

After the Second World War, several nationalist movements seeking to overthrow French colonial emerged in Indochina. Although the French army attempted to suppress these resistance movements, it was dealt a humiliating defeat at the *Battle of Dien Bien Phu* on May 7, 1954, approximately five months before the rebellion erupted in Algeria. After the defeat at Dien Bien Phu, Prime Minister Mendes-France arranged for as graceful a diplomatic and military withdrawal from Indochina as was possible under the circumstances. The resulting agreement between France, Vietnam, Laos, and Cambodia was the *Geneva Accords,* signed on July 21, 1954.

Significantly, as far as Algeria was concerned, during this same period the French government, under threat of open rebellion, was forced to grant important concessions to two of Algeria's neighbors, Morocco and Tunisia, regarding these countries' bid for independence. Although this conciliatory move was intended as a preemptive strike to prevent another armed conflict—"another Vietnam"—it only encouraged the Algerian rebels to push for their own independence more aggressively. French colonial rule was on the verge of collapse.

The imminent collapse of colonial rule must have been the prime minister's realization on that morning of November 12, as he stood before the National Assembly. Nevertheless, he maintained a never-say-die attitude:

One does not compromise when it comes to defending the eternal peace of the nation, the unity and integrity of the Republic. The Algerian departments [administrative districts of France] are part of the French Republic. They have been French for a long time, and they are irrevocably French. Between them and metropolitan France there can be no conceivable secession.

Again, the French government was in for a devastating surprise. But only gradually was it fully recognized in Paris that the All Saints' Day insurrection signaled the beginning of an unyielding liberation struggle against French colonial rule in Algeria.

The war of independence raged on for another eight years. Then, on July 1, 1962, during a cease-fire, 6 million Algerians cast their ballots in a referendum on whether or not they wanted their independence from France. The vote for independence was virtually unanimous. Two days later, on July 3, 1962, Premier Charles De Gaulle pronounced Algeria an independent country. The people of Algeria, however, preferred July 5, 1962, as their day of national independence—it was the 132nd anniversary of the French invasion and occupation of Algeria.

The Algerian struggle for liberation had been long and costly. Nearly eight years of revolution had cost anywhere from 300,000 (FLN estimate) to 1 million (Algerian government estimate) lives. In addition, French military authorities listed their losses at nearly 18,000 dead, 6,000 deaths being from non-combat-related causes, and 65,000 wounded. The struggle was costly to Europe as well: European civilian causalities exceeded 10,000.

Historians speculate that the actual figure of war dead and wounded was surely far in excess of the original FLN and official French estimate, though perhaps not as high as 1 million. Certainly, thousands of Muslim civilians lost their lives in French army bombing raids and vigilante reprisals. The war uprooted more than 2 million Algerians who were forced to relocate in French concentration camps if they did not flee to Morocco, Tunisia, or into the distant Algerian wilderness, where many thousands died of starvation, disease, and expo-

Algeria

Berber Village, c. 1890 *This Berber village is in Djurdjura, the mountain coastal region of northern Algeria. It is a harsh area in which to live. Bare ridges and deep gorges of swirling streams offer little pasture land for the nomadic flocks. However, population density is extremely high, and most villages are overcrowded. Today, Berbers makeup about 15 percent of the population—and nearly two-thirds of the nearly 5 million Berbers live in the Djurdjura area.*

sure. More thousands simply vanished. By any estimation, the price for Algerian freedom was high.

Today, as a result of its hard-won and costly freedom, Algeria has become a symbol of postcolonial independence. It is a nation that waged a highly visible war against a European colonial power, in the middle of the twentieth century, and won! That triumph tells a lot about the country and the determination of its people, but it by no means tells the complete story. For that we will have to journey back into time a few thousand years.

At the dawn of recorded history, the region later known as the Maghrib, or "the place where the sun sets," was inhabited by the

Berbers, a people who, as mentioned before, referred to themselves as *Imazighen* or "Free People." The native inhabitants of modern-day Algeria claim descent from those indigenous people.

The first inhabitants of the coastal area of present-day Algeria shared in an early Neolithic culture that was common to the entire Mediterranean coast before the fifteenth century B.C. The Neolithic period, or the New Stone Age, was the final stage of cultural evolution and technological development among prehistoric humans. It was characterized by stone tools shaped by polishing or grinding, dependence on domesticated plants and animals, settlement in permanent villages, and the appearance of such crafts as pottery and weaving.

These innovations spread from the Middle East across Egypt and into North Africa. South of the Atlas Mountains nomadic hunters and herders roamed a vast savanna, or tropical grassland, that eight thousand years ago stretched across what is now the Sahara. The "savanna people" left a vivid portrait of the world in which they lived in a series of rock paintings like those discovered at Tassili N'Ajjer in Algeria, one of the most famous sites of rock painting in North Africa. The images depicted in these paintings capture a lush Sahara overflowing with green, growing plants—a stark contrast to the arid desert it has since become. Painted over a period of 4,000 years, these paintings are the most complete existing record of a prehistoric African culture and are among the most remarkable stone-age remains to be found anywhere.

The culture depicted in the rock paintings of Tassil N'ajjer flourished until 4000 B.C., at which time the region began to dry up as a result of changes in the climate. Despite that climatic dessication, the prehistoric savanna is still believed to have given birth to the subsequent African civilization. The savanna people migrated in various other directions as well, throughout the region of North Africa, always trying to stay a few steps ahead of the encroaching desert. It is believed that these migrating peoples of North Africa merged over time into the distinct indigenous population that came to be known as the Berbers.

The arrival of Phoenician, or *Punic,* traders about 1100 B.C. brought the region into recorded history. Gradually, Punic trading posts were established along the entire North African Coast,

Algeria

Caravan, Algeria, c. 1880
This unusual photograph probably was taken near An Salah, an oasis town in central Algeria, the crossing point of trans-Saharan caravans which linked northern and central Africa. Note the length of this caravan. A caravan is a group of merchants or travellers journeying together for mutual protection in deserts. The camel is the common animal used because of its ability to go without water for days. And in hot water, on long journeys, a camel can carry about 350 pounds.

where the merchants of *Tyre* and, later, Carthage traded with the Berbers and paid them tribute to ensure their cooperation.

By the fifth century B.C., Carthage, the greatest of the Punic colonies, had extended its hegemony across much of North Africa. Carthage is located on the Bay of Tunis, northeast of modern Tunis, the capital and largest city of Tunisia. The rulers of Carthage cultivated good relations with the Berbers and developed defensive alliances with various kinship groups in the hinterland.

Being essentially a maritime power, Carthage hired Berber mercenaries for its overseas military expeditions, but the emerging power of Rome contested Carthaginian expansion in the western Mediterranean. Defeated in the long Punic wars in the third century B.C., Carthage was reduced by Rome to the status of a small and vulnerable African state at the mercy of the various Berber ethnic groups. Yet the influence of the seafaring Punic civilization remained deep-rooted in North Africa.

The basic unit of social and political organization among the Berbers was the extended family, usually identified with a particular village or traditional grazing grounds. Families in turn were bound together in the clan. An alliance of clans, often tracing their origins to a common ancestor as a symbol of unity, formed kinship groups. For mutual defense these kindred groups formed confederations that, because war was always a constant threat, became in time a permanent way of life.

Some Berber leaders, because they were successful in battle, were able to establish temporary territorial states by imposing their rule on both defeated enemies and allies who sought continued security and protection of a strong leader and warrior. But these alliances were easily fragmented, and the dynasties these warrior-kings sought to establish rarely lasted a generation. Nevertheless, by the second century several large though loosely administered Berber kingdoms had emerged, supported by farmers and villagers who looked to the kings to protect them from the raids of nomadic warriors and hunters. Two such kingdoms were established in *Numidia,* the boundaries of which corresponded roughly to those of modern-day Algeria. The kings of Numidia ruled in the shadow of Carthage and later Rome. After Carthage was vanquished, Berber kings formed strategic alliances with factions vying for power during the Roman civil wars of the first century B.C.

One of the most illustrious of these Berber kings was Masinissa, (240–148 B.C.) who allied his people with the Carthaginians in Spain. When circumstance demanded, Masinissa shifted his support to Rome just before the fall of Carthage in 202 B.C. With Roman patronage he united Numidia and extended his authority from Moulouya to Cyrenaica. The royal house of Masinissa continued to rule Numidia until after his death, at which time the kingdom was divided among his heirs, with representatives of the Roman Empire making all of the important decisions.

In 118 B.C., Roman hegemony in North Africa was challenged by Jugurtha, Masinissa's grandson. Like his grandfather, he sought to unite the Berber kingdoms, which brought him into direct conflict with Roman authority. At the end of a long and

Algeria

costly war, Jugurtha was defeated, imprisoned, and deported to the Roman capitol where, according to tradition, he died of starvation. After Jugurtha's death the Romans continued to dominate Numidia. They sought to control only those areas that were economically useful or that could be defended without too much manpower. Called the "granary of the empire," the North African provinces were valued also for their agricultural exports, which were Rome's principal source of food. Slave labor was common, but on the vast imperial estates owned by the Roman elite, land was rented to Berber tenants, who paid rent and taxes in the form of labor and grain that went to feed the Roman army and provide free bread for the poor in Rome.

Although, without a doubt, the Roman Empire profited from its relationship with the Berber people of North Africa, at the same time, Numidia prospered under Roman rule. The ruins of Roman cities seen in present-day Algeria attest to the civic and economic vitality of North African provinces, where even the smaller towns enjoyed the amenities of urban life found in every corner of the Roman Empire. In time, these towns grew into miniature Roman cities. This sense of prosperity continued until the fall of the Roman Empire.

In the meantime, by the end of the fourth century, the settled areas of North Africa had been Christianized, and inroads had been made as well among the Berbers who lived farther into the interior and sometimes converted en masse. As Christianity was making its many inroads, there were also schismatic and heretical movements like *Donatism* that developed within it, often as forms of political protest. Under the leadership of Aelius Donatus, the Donatists split from the Church of Rome to form a more "pure church." This populist movement at odds with Roman imperialism gathered momentum with the Berber popluation. "You come with edicts of emperors; we hold nothing in our hands but volumes of scriptures," the Donatists observed. They insisted that ministers who dispensed the sacraments be "sinless," and that their church be composed of people from North Africa, selected from "a small body of the chosen."

Donatism posed a threat not only to Roman Christian authority but to Roman legal and civil authority as well, because what began as a call for a separate church soon became a movement of Berber resistance against Roman power in North Africa. Donatism has been viewed as an early precedent for the anti-imperialist and nationalist sentiment that later fueled the Algerian War of Independence. It was a declaration of Berber identity as well. Some historians have gone so far as to characterize the Donatists as the urban poor and rural peasants revolting against the land-owning class of the Roman church. Interestingly, it was one of the first resistance movements in which women played an active part.

It was against the Donatist threat to the African church that Saint Augustine (A.D. 354–430), bishop of Hippo Regius (Annaba), directed his sermons and books, including his autobiographical *Confessions,* which merited him recognition as one of the Latin Fathers of the Church. Augustine was born on November 13, 354, in Tagaste, Numidia, now Souk-Ahras, Algeria. Although his work affected Western Europe more than it did North Africa, Augustine played a significant role in the Roman Church's attempts to suppress the Donatist movement. He died on August 28, 430, a day that Catholics continue to remember in his honor.

According to tradition, as St. Augustine lay near death, he heard the clamor of weeping and wailing, and when he asked what all the noise was about, he was informed that the barbarians were at the gate. The barbarians that were clamoring at Hippo's gate were *Vandals,* a name that has come down to us as synonymous with willful destruction and lawlessness. They were a Germanic ethnic group that, fleeing the Huns, traveled westward into Spain and then eastward again along the North African Coast of the Mediterranean. Conquering Hippo and the surrounding area at the time of the death of St. Augustine, they remained the lords of the region for a century until they were subdued in turn by the Byzantine general Belisarius, after which they disappeared from history.

The Vandals may have been "vandals," but they were also Romans and Christians, too. The problem was that they were *Arians,* meaning they did not believe in the divinity of Jesus Christ. Arianism was a Christian heresy of the fourth century that denied the full divinity of Christ. It was named for its founder, Arius, a native of Libya. The Arians believed that Jesus, the Son of God, was only like God and was not of one spirit, or substance, with God. This distinction was of supreme importance in the fourth and fifth century, and even has survived in one form or another to this day. For example, Jehovah's Witnesses assert that Arius was an ancestor of their founder, Charles Taze Russell.

All such distinctions and controversies became inconsequential, especially in North Africa, when the irresistible force of Islam exploded out of Arabia, first into Persia and Syria, then into Egypt, and then westward across the Maghrib and then into Spain. By the beginning of the eighth century Muslims controlled all of North Africa, and the Roman Empire slowly but irrevocably faded away.

During the early seventh century, a trader named Muhammad (570–632) was meditating in a cave near Mecca in what is now Saudi Arabia when he experienced a vision of the archangel Gabriel, who declared him to be a prophet of God. Other revelations followed, and Muhammad began to preach to others, reciting in verse the instructions he had received from God. These revelations became the *Qu'ran,* or Koran, the sacred scripture of Islam.

The Prophet Muhammad was also known as *al-Amin ir,* "The Trusted One," and his followers were known as *Muslims,* or "ones who surrender to God." By the time of the Prophet's death in 632, most of the various ethnic groups and towns of the Arabian Peninsula had come under the banner of the new monotheistic religion of *Islam* (an Arabic word that means submission or surrender). Viewed as uniting the individual, the state, and the society under the omnipotent will of God, Islamic rulers exercised both temporal and religious authority. Adherents of Islam collectively formed the *Dar al Islam,* or *"House of Islam."*

Within a decade of Muhammad's death, Arab conquests together with immigration and trade had carried Islam north and east of Arabia. After completing their conquest of Egypt in 642, Arabs began a steady advance into territories of North

Africa inhabited by the Berbers. They called these territories *Bilad al-Maghrib,* "Land of Sunsets," or simply *Maghrib.* Encountering militant resistance from the Berbers slowed their advance for several years, and efforts at permanent conquest were resumed only when it became apparent the Maghrib would be of strategic importance in the Muslims' campaign against the Byzantine Empire.

In 670 Arab armies completed their conquest of the Roman province of Africa—*Ifiquiya* in Arabic—and founded their city of *Al Qayrawan,* or *Kairovan,* in northern Tunisia. This town, one of the holy cities of Islam (it is called "the City of One Hundred Mosques) lies on the Low Steppes, a semiarid alluvial plain southeast of the Central Tell. Founded on the site of the Byzantine fortress of Kamouninia, it served as the camp from which the offensive was launched that resulted in the Islamic political and religious conquest of the Maghrib. Al-Qayrawan was chosen as the capital of the Maghrib by the first Aghlabid ruler in about 800. It also served as the political center through the Fatimid and Zirid dynasties into the eleventh century, becoming one of the great administrative, commercial, religious, and intellectual centers of Islam. Finally, in the fifteenth century it became the administrative center under the Almoravids. The French occupied it in 1881.

The Arabic language, which until the conquests was confined to the Arabian Peninsula, spread in the region with Islam. In the lands from Iraq to Algeria the populations became essentially Arabic-speaking. Other languages, such as Greek, Aramaic, and Coptic, steadily disappeared from common use, surviving mostly only in liturgy and religious writings.

The possible exception to this rule were the various dialects spoken by the Berber people of North Africa. The extent to which the Berber culture and language were influenced and changed by Arabic culture, particularly in Algeria, is a question still debated by historians. One of the defining characteristics of the Berber people is their love of independence and their impassioned religious temperament, so a question remains as to whether Islam shaped their language and culture or whether they appropriated and shaped Islam in their own image. Cer-

Algeria

tainly, many Berber confederacies violently resisted Arab rule, at least initially. The revolt led by the Berber Queen Kahina, (*al-Kahina,* "The Priestess") in the seventh century is one of the best known examples. On the other hand, other Berber confederacies entered into strategic alliances with Islamic rulers that led to the gradual conversion of most Berbers and their increasing use of the Arabic langauge. So although the Berbers' experiences with Islam have varied widely, it may be said that Berber tradition remade Islam in North Africa just as Islam remade the Berbers.

The North African Islamic dynasties, including the *Almoravids, Almohads,* and the *Marinids,* were all led by Berbers, and some incorporated elements of Berber confederate leadership, such as succession based on female kinship. As these Islamic empires conquered portions of southern Spain, Berber influence traveled throughout the Mediterranean.

In the meantime, following Berber conversion to Islam, Algeria became a province of the *Umayyad Caliphate.* The Umayyad, also *Omayyad,* was the first great Arab Muslim dynasty of *caliphs,* the religious and secular leaders founded by Muawiyah I in 661 and lasting until 750. Under the Umayyad dynasty, political and social power remained in the hands of a few elite families from Mecca and Medina. As a result, the Muslim community, which had grown considerably as the empire expanded, became discontented, especially under the Umayyads.

In Algeria, this internal conflict and a disagreement over the succession of the caliphate enabled the Berbers to form their own Islamic government in the eighth century. Many of them joined the branch of Islam known as *Shia,* and they founded several related kingdoms. One of the most important was *Rustamids at Tahert* in central Algeria. Tahert prospered in the eighth and ninth centuries, but between the eleventh and thirteenth centuries two successive Berber dynasties, the Almoravids and the Almohads, brought northwest Africa and southern Spain under a single authority. *Tiemcen,* the capital under the Almohods, became a city of fine mosques and schools of Islamic learning, as well as a handicrafts center. Algerian seaports like Bejaia, Annaba, and the growing town of

Algiers, or *Al Jazair,* carried on substantial trade with European cities, supplying the famed Barbary horses and fine leather and fabrics to European markets.

The collapse of the Almohads in 1269 ignited a fierce trade war between vying Mediterranean seaports, both Christian and Muslim. In 1510 Spain captured and fortified a strategic harbor known as the Penon. For the next eight years Spain controlled the flow of trade and the profits. In 1518 Algiers proclaimed itself part of the Ottoman Empire, and the Spanish were driven out. While ruled by the Ottomans, Algiers became the capital of the infamous Barbary Coast. For the next 300 years, Barbary pirates preyed on European and later American ships. European states paid tribute regularly to ensure protection for their ships, and prisoner ransom brought in a rich income to the province.

In 1815 the American naval captain Stephen Decatur led an expedition against Algiers, forcing its governor to sign a peace treaty promising to end attacks on U.S. ships. Nevertheless, the piracy continued until 1816, when combined Dutch and British navies almost completely destroyed the Algerian fleet. Even after suffering this devastating defeat, Algiers remained a pirate port until 1830, when the French invaded it and made the entire country of Algeria part of its colonial empire, a situation that continued for the next 132 years.

France formally annexed Algeria in 1834, against fierce resistance from the various Berber kinship groups and confederations that had grown accustomed to indirect Ottoman rule. It was not long before a charismatic leader emerged and united the people of Algeria in revolt against French colonial rule. His name was Abd al-Qadir, an Islamic "holy man" and military leader who claimed descent from the Prophet Muhammad.

Still considered a hero of anticolonial resistance by many contemporary Algerians, Abd al-Qadir created an Arab-Berber alliance to oppose French expansion throughout the 1830s and 1840s. He also organized an Islamic state that at one point controlled the western two-thirds in the surrounding Oran province of Algeria. His ability to unite Arabs and Berbers who had been enemies for centuries owed in part to the legacy of his father, the leader of the Hashim confederation in Mousadar (Mascara) and

Algeria

Street Scene, Algiers c. 1890
In 1830 France conquered Algeria. This ended three centuries of Algerian history as an autonomous province of the Ottoman Empire. Algeria obtained its independence from France 132 years later. Algiers, the capital and chief seaport of Algeria, is the political, economic, and cultural center of the nation.

head of a *Sufi Muslim* brotherhood. The term *Sufi,* or "man of wool," was coined in the early ninth century for mystics whose ascetic practices included wearing coarse woolen garments—*sufu*. These new Sufi orders often helped African Muslims respond to the tumultuous changes brought about by colonialism.

In 1826 Abd al-Qadir and his father made a pilgrimage, or *hajj,* to Mecca in Saudi Arabia, the birthplace of the Prophet Muhammad. Upon his return, Abd al-Qadir's reputation as an Islamic religious leader grew, and Arabs and Berbers looked to him to lead the resistance against the French.

In Algeria, Abd al-Qadir's legacy remained an inspiration throughout the Algerian War of Independence (1954–1962). In 1968 the newly independent nation erected a monument to Abd al-Qadir in the place where a French monument to General Bugeaud had stood, and took up his green and white standard as its flag.

Main Entrance, Medersa Bou Inania, Fez, c. 1890 *A* medersa *is a residential Islamic college for learning the Koran. Medersa Bou Inania was constructed by Sultan Abou Inan at some time between 1351 and 1357. It was the largest constructed throughout Morocco. This type of architecture used precious woods, bronze, marble, and onyx decorations. It shows the exquisite taste, creativity, and wealth of this period of Moroccan history.*

2

Morocco

While Algeria was caught in the throes of violent revolution, Morocco, its neighbor to the north, was experiencing a rebirth of its own, but it was a different, quieter, less violent, and less costly kind of revolution. On March 2, 1956, after seven minutes of formal negotiation in the *Salon d' Horloge* of the French Foreign Ministry, forty-four years of French colonial rule in Morocco ceased. Christian Pineau, the French foreign minister, and Si M'Barek Bekkai, premier of Morocco's new transitional government, initialed the new negotiated declaration of independence:

> As a result of Moroccan advances on the path of progress, the "treaty of Fez" of March 30, 1912, no longer corresponds to the necessities of modern life and can no longer govern Franco-Moroccan relations. In consequence, the Government of the French Republic solemnly affirms its recognition of the independence of Morocco, which in particular implies its right to diplomatic representation and the maintenance of armed forces, as well as its desire to respect and have respected the integrity of Moroccan territory, guaranteed by international treaties. The negotiations that have just begun in Paris between

Modern map of Morocco

Morocco and France, sovereign and equal states, have as their object the conclusion of new agreements that will define the interdependence of the two countries in the fields in which their interests are in common, that will thus organize their cooperation on the basis of Liberty and equality, notably in matters of defense, external relations, economy, and culture, and that will guarantee the

MOROCCO

rights and freedoms of the French settled in Morocco and of the Moroccans settled in France, subject to respect for the sovereignty of the two states.

One month and five days later, Si Bekkai conducted the first significant affairs of the independent and sovereign State of Morocco. With the Spanish foreign minister, Martin-Artajo, he signed an agreement that ended Spanish control over the northern zone of the Sharifian Empire. This act of reunification made Morocco the first country in North Africa to liberate itself from French colonial rule and take its place in the family of independent and sovereign nations of the world.

Morocco, officially the "Kingdom of Morocco," *Al-Mam-Lakah al-Magribiyah,* or simply *Al-Maghrib,* is both an ancient kingdom and a new nation. It is, along with Ethiopia, one of the world's oldest states by virtue of its continuous 1,200-year history as a political entity. Its problems are closely identified with those of the developing nations of Africa, the Middle East, and Latin America. Morocco's cultural traditions, social customs, religious values, and political institutions—even the very fabric of its cities, like Casablanca, Rabat, Fez, Marrakesh, and Tangier, to name a few—all testify to the great diversity of life and heritage intricately woven like unbroken cords throughout the country's long and varied history.

Because of its lack of natural harbors, its rugged mountainous interior, and its distance from imperial and colonial powers, Morocco remained relatively free from foreign invasion until the early twentieth century. Consequently, the people have preserved their proud traditional character, a rich and diverse blend of Berber, Arab-Islamic, Iberian, and African culture. Those elements were interwoven with European cultural influences during the colonial period of 1912–1956, when Morocco was a protectorate of both France and Spain.

The people of Morocco take great pride in their country's distinctiveness and in its historical role as the heartland of empires that once encompassed much of North Africa and Spain. Morocco has been the geographical and political bridge as well as the cultural liaison for the Arab-Islamic world with

Kaid of Shaonia, 1896 *Sultan Moulay Hassan I (1873-1894) attempted to modernize Morocco while keeping the nation independent of European colonialism. However, his goal seemed almost impossible. For example, the students he sent to Europe to learn the latest medical and technological skills returned unable to cope with the political realities of Morocco. The various trade treaties made in the 1850s had placed European merchants and their numerous Moroccan agents outside the sultan's authority. Between 1844 and 1873, the national currency had lost 90 percent of its value. Commerce was conducted in French francs. Foreign consuls in the ports of Tangier and Casablanca even constructed lighthouses and public works—and maintained a separate postal system.*

Moulay Hassan's two sons presided over the last stormy decade of Moroccan independence before the imposition of French rule in 1904. But the royal court now was composed of European adventures and unscrupulous advisers. Popular discontent became common. Meanwhile, tribal chiefs and administrators (kaids) appointed by Moulay Hassan, notably in the High Atlas Mountains, expanded their power in the vacuum created by the chaotic central government.

In 1896 Great Britian sent Sir Arthur Nicholson to deal with the High Atlas mountain kaids in an attempt to impose unity in Morocco. This is a photograph of the leading kaid, taken by Nicholson's assistant.

Morocco

Europe and West Africa. Although Morocco shares a kindred spirit with all the countries of the Maghrib, throughout the past twelve centuries it has been Islam more than any other influence that has given the Moroccan people a common identification with those other countries of the Maghrib. Other aspects of its culture are distinct unto itself. Something else that sets Morocco apart from its brothers and sisters of the Maghrib is the fact that it is the only country in North Africa that escaped the influence of the Ottoman Empire, and its brief period of colonial occupation did little to alter the country's long tradition of independence. If anything, the colonial period was when a genuine Moroccan nationalism came into being.

The Kingdom of Morocco is a constitutional monarchy, in which the powers of the hereditary king are restricted to those granted under the constitution and laws of the nation. The government is divided into two separate branches, the executive, represented by the offices of the Chief of State and the Head of Government. King Mohammed VI is the present chief of state of Morocco. He took the throne on July 23, 1999. Prime Minister Abderrahmane Youssouf fulfills the duties of the head of government. The other branch of government is the legislative, consisting of a bicameral parliament. The upper house, the Chamber of Counselors, has 270 seats, and all the members are elected indirectly by local councils, professional organizations, and labor syndicates for nine-year terms. One-third of the members have their terms renewed every three years. The lower house, the Chamber of Representatives, consists of 325 seats, and members are elected by popular vote for five-year terms.

Morocco is the only country in Africa that borders both the Atlantic Ocean and the Mediterranean Sea. Located in northwestern North Africa, it covers a total landmass of 175,186 square miles, or 453,730 square kilometers. Slightly larger than the state of California, it lies directly across the Strait of Gibraltar from Spain. It borders Algeria to the east and southeast, Western Sahara to the south, the Atlantic Ocean to the west, and the Mediterranean Sea to the north.

Morocco has the broadest plains and the highest mountains in North Africa. Like Algeria, it can be divided into several distinct geographical regions. There are the *Er Rif,* "highlands," which parallel the Mediterranean coast; the Atlas Mountains, which extend across the country between the Atlantic Ocean and the Er Rif; the *Taza Depression,* a region of broad coastal plains along the Atlantic Ocean; and finally, the plains and valleys south of the Atlas Mountains, which merge with the Sahara along the southeastern borders of the country. Most Moroccans inhabit the Atlantic coastal plain.

The coastal regions of present-day Morocco shared in the early Neolithic culture that was common to the entire Mediterranean littoral. Archaeological remains suggest that 8,000 years ago, south of the great mountain ranges in what is now the Sahara Desert, a vast savanna, abundant with game, supported Neolithic hunters and herders. This culture flourished until the region began to desiccate as a result of climatic changes after 4000 B.C.

It is believed that the Berbers entered Moroccan history toward the end of the second millennium B.C., when they made initial contact with "oasis dwellers" on the steppe, who may have been the last remnants of the earlier savanna people. Linguistic evidence resulting from the study of Berber languages indicates that southwestern Asia was the point from which some of the ancestors of the Berbers began their movement into North Africa. An Egyptian inscription dating from the Old Kingdom (2686–2181 B.C.) may be the earliest recorded testimony of their westward migration, which over successive generations extended their range from Siwa in eastern Egypt to the Niger River basin.

The term *Berber* refers to any of the descendants of the pre-Arab-Islamic inhabitants of North Africa. Through the centuries they have intermarried with so many other ethnic groups, most notably Arabs, that Berbers are now usually identified primarily on a *linguistic* rather than a strictly ethnic basis. The Berber language is a branch of the Afro-Asiatic, formerly Hamito-Semitic, language family and comprises about 300 related local dialects. The other branches of the Berber language family include Egyptian, Semitic, Cushitic, and Chadic.

Morocco

Berber Woman, c. 1925
This Berber woman was photographed in the small village of Asni in the High Atlas Mountains of Morocco. In the late twentieth century there were about 9 million Berbers living in Morocco. The simplest Berber political structure is found in the High Atlas area where all adult men meet in the village square to make community policy. Women cannot participate, but within Berber society, women do have a great deal of personal freedom.

Under the French protectorate (1912–1956), Arabs and Berbers were perceived as two distinct ethnic groups. The Arabs were considered an urban population loyal to the sultan of Morocco. The Berbers, regarded by the French as a rural, tribal people, were considered outside the sultan's authority. This French assumption that Berbers and Arabs were culturally distinct groups was connected with their colonial policy of divde and rule.

However, many Berbers speak Arabic; and those Arabs living near Berber villages in the mountains often learn the Berber language. Ethnic identification has been further blurred by the sharing of Islam.

The Berber languages are spoken in scattered areas throughout North Africa, from Egypt westward to the Atlantic Ocean, and from the Niger River northward to the Mediterranean Sea. In total, about 11 million people speak Berber languages. The more important dialects are Tamashek, or Tuareg, in the central Sahara and south of the Niger; Tashelhayt, or shilha, in Morocco and Mauritania; and Zenaga in Mauritania and northern Senegal.

Tuareg, c. 1880 *This posed studio photograph is of a Tuareg, a Berber-speaking nomadic herder who lives in the Sahara Desert. He is holding a traditional weapon—a combination sword and iron lace. Henri Duveyrier (1840–1892), the French explorer, was among the first Euopeans to study the Tuareg people, with whom he lived for months at a time. Although his book,* Exploration of the Sahara, *was written more than 100 years ago, it still ranks among the great adventure stories about the Sahara.*

Morocco

Tuareg Chief, Southern Libya, c. 1910   *The Northern Tuaregs live in the Sahara Desert while the southern Tuaregs live in the savanna areas of Niger and Mali. The Tuareg do not recognize national boundaries. Sir Hanns Vischner (1876–1945) described in* Across the Sahara from Tripoli to Bornu *(1910) as Asben caravan Tuaregs from Mali arriving at the oasis in southern Libya of the chief in this photograph:*

It was an Asben caravan of over 8,000 camels and 1,000 men. The arrival of these caravans is naturally the most important event of the year to the people of the oasis. The Asbenawas bring millet and grass from Air [Ar Massf region of central Mali], wood for camel saddles, Manchester cloth, Hausa robes, and all the luxuries which can be found in the Kano [Nigeria] market. Men, women, and children arrive from all the villages [on the oasis] to buy their provisions for the year, which the [Absens] give them in exchange for alst and dates.

The great encampment, with the many thousand camels, the stacks of grass, and piled-up loads, looked like an immense fair. The oasis, of course, could never support all these animals, so the Asbenawas, before they leave Air, feed up their camels on the fattest grazing grounds, and then, having chosen only the fittest animals, load about a third of the number with grass, which is used for fodder on the way. Great quantities are buried in the sand at intervals, to be used on the return journey; for on that desolate stretch of desert, water is very rare and not a blade of grass grows between the interminable sand dunes.

Not much is known about the ancient Libyan language, also called Numidian. Some evidence of its existence has been discovered in inscriptions found in Tunisia and Algeria dating from the times of the Roman Empire and written in a consonantal quasi-alphabetic script that still survives in a modified form among the Tuaregs of Sahara. Part of the problem with tracing the origins of Berber languages and particular dialects is that it is primarily a spoken language and its written form is little known and rarely used.

Although Berbers left no written records, they were well-known to writers of classical antiquity. The earliest known reference to Berbers comes from Hecateus, writing in the sixth century B.C. He refers to them as "Libyans." Herodotus and Polybius made reference to them as well, and Sallust's description of the Berber way of life, written in the first century B.C., provides details and other information still useful to historians.

Phoenician traders who explored the western Mediterranean before the twelfth century B.C. set up depots for the mining of salt and ore along the coast and up the rivers of the territory that is present-day Morocco. Tangier, Te'touan, Melilla, Essacuira (also called Mogador), and Ceuta all had their origins as Punic trading posts, where merchants of Tyre, Sidon, and later Carthage developed commercial relationships with the Berber kinship groups of the interior and paid them an annual tribute to ensure their cooperation. By the fifth century B.C., Carthage, the greatest of the overseas Punic colonies, had extended its hegemony across much of North Africa. Punic settlers on the Atlantic coast bartered their merchandise for gold from the western Sudan region, in quest of which the Carthaginian Admiral Hanno made his fabled voyage to the mouth of the Senegal River.

In spite of all the industry and cultural exchange taking place on the coastal regions, Berbers, who remained beyond the reach and lure of the commercial enclaves of their Punic visitors, continued to live their lives remarkably the same as they had done for generations. If there was an exception to this rule, it was in the nature of religion. Berbers displayed a fierce loyalty to their

Morocco

local gods, and their worship was exceedingly personal and enthusiastic. But Berbers also demonstrated a surprising gift for cultural assimilation, readily synthesizing Punic cults and religious practices, and interweaving them into their nature worship, magic, and reverence for holy places. They would, in later years, do the same with Greco-Roman and Egyptian deities, Judaism, Christianity, and finally Islam—although, it must be admitted that Islam exerted a more powerful and longer lasting influence than the other religions.

Assimilating and reshaping other people's culture to fit the image of the world they had created for themselves was not the Berbers' only means of resistance. For mutual protection and defense, kindred ethnic groups joined in confederations, which, because of the constant threat of warfare, became an almost permanent way of life. One of the most powerful confederations was under the leadership of a Berber king by the name of Juba II. He was the son of Juba I, the king of Numidia, who sided with the followers of Pompey, and the Roman Senate in their war against Julius Caesar in North Africa, (49–45 B.C.). Pompey died the following year, but African resistance continued under Metellus Scipio, to whom Juba I was also allied.

In 46 B.C., Caesar himself came to subdue them. Juba I had to divide his substantial army of infantry, cavalry, and elephants because his kingdom had been invaded from the west by Caesar's allies, Bocchus, king of Mauritania, and an Italian adventurer by the name of Publius Sittius. Juba I was defeated along with the other supporters of Pompey at the Battle of Thapsus, and his general in the west was killed by Sittius. Rather than face the disgrace of this terrible defeat, Juba I committed suicide. His suicide left the fate of his son in the hands of Juba's enemies.

As a child of five, Juba II was paraded in Rome in Caesar's triumphant procession after the death of Juba I, but subsequently was treated as a member of the court and given a good education in Italy. Octavian, the future emperor Augustus, befriended Juba when he was a young man and in 29 B.C. made him ruler of his father's former kingdom of Numidia, which had become a Roman province after the death of Juba I in 46

B.C. In 25 B.C., Juba also became ruler of Mauritania, which he governed until his death in A.D. 24. His first wife was Cleopatra Selene, daughter of the famed Cleopatra and Mark Antony. She exercised great influence on his policies. They also had a son, Ptolemy, who would rule Mauritania after his father's death. Juba was a prolific writer in Greek on a variety of subjects, including history, geography, grammar, and the theater.

As evidenced by these few brief passages, the history of the region constituting present-day Morocco has been shaped by the interaction of the indigenous Berber population and the various peoples who successively invaded the country. The first of these foreign invaders were the Phoenicians who, in the twelfth century B.C. established trading posts along the Mediterranean coast. These colonies were later taken over and extended by the Carthaginians. The conquest of Carthage by Rome in the second century B.C. led to Roman domination of the Mediterranean Coast of North Africa.

Even though Roman authority and influence was extensive in regions controlled by the Empire, it must be kept in mind that Rome expanded its authority only to those areas that were economically useful or that could be defended without additional manpower. As a result, Roman administration never exerted much authority or influence beyond the restricted regions of the coastal plains and valleys. When Rome did attempted to do so, it was faced with immediate, continual, and fierce opposition, leading one Roman official to remark, "These people can be conquered but not subjugated." Indeed, the people of Morocco fought long and hard to maintain their independence; nevertheless, it was the Vandals and not Berber resistance that finally brought an end to Roman domination of North Africa.

The Vandals were a Germanic people who maintained a kingdom in North Africa from A.D. 429 to 534. They dominated what is now Algeria and northern Morocco by 435 and conquered Carthage in 439. With their rule firmly established, the Vandals eventually annexed Sardinia, Corsica, and Sicily, and their pirate fleets controlled much of the western Mediterranean. Under Gaiseric, their most notable leader, the Vandals

Morocco

Palace of Septimus, Leptis Magna, 1925 *In ancient times, the cities of Oea, Sabratha, and Leptis Magna formed the African Tripolis, or Tripoli. During the fifth century A.D. the vandals destroyed the walls of Sabratha and Leptis Magna. Destruction of those two cities resulted in the growth of Tripoli, which had previously been the least important of the three. The Romans controlled African Tripolis from 146 B.C. until about A.D. 450. This is a photograph of a pillar from the palace of Septimius, which was part of the ancient Greco-Roman amphitheater complex at Leptis Magna.*

even invaded Italy and captured Rome in June 455. For a fortnight they occupied the city and systematically plundered it, carrying off many valuable works of art. The Byzantine general Belisarius finally defeated the Vandals in 533. From that point forward they no longer played an important role in history.

Compared to the impact made by the previous foreign invaders, Islam was like an explosion that shook the Maghrib to its very foundation. After the arrival of Arab-Islamic culture the history of North Africa was changed forever. But it was not an abrupt, violent, revolutionary change. Quite the contrary, it was a gradual and methodical change that took place over many centuries and many generations. The independent and resistant nature of the Berber people had a great deal to do with the rate of change that took place in North Africa, particularly Morocco.

When Muhammad died in Medina, called *Medinat-en-Nabi,* or City of the Prophet, in June 632 after a painful illness, he left no instructions regarding who was going to succeed him. The Islamic community was not only in a state of grief and despair over the loss of their beloved Prophet, but his death had also brought confusion about the future of the community. Muhammad had no sons. His most trusted companions were Abu Bakr, father of his young wife Aisha, and a son-in-law, Ali, husband of his daughter Fatima. But there was no traditional or legal precedent governing the choice of either one of these men as the Prophet's successor. After a prolonged debate within the Muslim community, Abu Bakr was chosen as *Khalifat rasul-Allah,* "Successor to the Apostle of God." From this appellation came the term *caliph.*

The first task of the new caliph was to restore order and the sense of security that had been disrupted after the Prophet's death. For this responsibility he depended largely upon his chief marshal, Khalid ibn al-Walid, whose name meant "The Sword of Allah" and who was to become one of the greatest Arab generals. Although he fought against Muhammad at the *Battle of Uhud,* Khalid was later converted to Islam and joined Muhammad in the conquest of Mecca in 629. Thereafter he commanded a number of conquests and missions in the Arabian Peninsula. After the death of Muhammad, Khalid recaptured a number of provinces that had broken away from Islam. He was sent northeastward by Abu Bakr to invade Iraq, where he conquered Al-Hirah.

There were two primary motives for Islamic expansion, and neither one of them had anything to do with religion, at least not directly. One was to find an outlet for the martial energies of the Bedouin warriors who made up the bulk of the Islamic converts, and the other was to search for booty and supplies to sustain an impoverished Muslim community. The expansion began tentatively with a series of probing raids across the borders into Mesopotamia. Initially, as already stated, the Arab raiders' primary objective was the spoils of battle, usually in the form of ransom and booty. But once they discovered that resistance to their schemes was weak or nonexistent, and that they even gained recruits for their campaigns along the way, their ambitions became increasingly comprehensive. In this frame of mind their conquest of the Sassanid Empire of Persia and the Byzantine Empire, the two great empires of the Middle East, was accomplished with astonishing speed and devastation.

Again, the initial driving force behind the Arab-Islamic explosion across Arabia was less religious zeal than the pressures of hunger and want bearing down on a people of a barren land. But as the spark of conquest erupted into flame, the nature of the expansion inevitably changed. A great empire was being born.

The first real setback for the Arabs came in North Africa at the hands of the Berbers, who according to Peter Mansfield in *The Arabs*, "were no luxury-loving city-dwellers but warlike nomads like themselves." It took nearly a century for the Arabs to subdue the Berbers of North Africa. But the interaction was ultimately more influential and the changes more thorough than any other foreign invaders had achieved, especially the Romans. Christianity was replaced by Islam and Latin by Arabic in all the Romanized areas of North Africa. The Latin and Greek population of the cities withdrew to Spain and Sicily. The Berber population, on the other hand, retained many of their customs as well as their language and dialects, even after they had embraced Islam and intermarried with the Arabs.

Street Scene, Fez, c. 1923 *Fez, the oldest of Morocco's imperial cities, is an important religious, intellectual, and cultural center. It is also renowned for its traditional crafts.*

Morocco

Potter's Shop, Fez, c. 1923   *Arab cities were designed with the souk, or market, as the center of the community. For example, the streets around a grand mosque are usually the busiest, as the outer walls are commonly obscured by workshops and stalls whose rents help pay for the upkeep of the building. The souks of Fez are grouped according to speciality.*

Washing Clothes, Fez, c. 1923 *These people are washing on the banks of the Wadi Fez. The Wadi joins the Sebou River, which empties into the Atlantic Ocean.*

Most Berbers did not convert to Islam until Musa ibn Nusayr pushed west across the border into the region the Arabs call *Ifriqiya*. Morocco then fell under the political and religious leadership of the Umayyad Dynasty, which was based in Damascus. As Berbers converted to Islam and intermarried with Arabs, they became a vital military force for Islamic expansion, especially in Spain.

The first Islamic dynasty of consequence to rule the nation of Morocco was named after its founder, Idris I. The *Idrisid Dynasty* was founded in 788 and lasted until the end of the tenth century. Idris I was the great-great-grandson of Fatima and Ali, the daughter and son-in-law of the Prophet Muhammad. According to one legend, Idris founded twin Arab and Berber cities at Fez (also called Fes). Another legend credits Idris with founding one city, and his son, Idris II, with founding its twin.

Morocco

With the opening of al Qarawiyin University in 859, the city of Fez flourished as a center of learning, attracting Muslims from southern Spain and Ifriqiya. Some historians believe that this was the beginning of the first Moroccan state, but it did not encompass the entire area of modern-day Morocco. Certain territories remained under the control of the Umayyad *emirs* in southern Spain or the Fatimid Empire, as well as several Berber confederacies that maintained autonomous rule in neighboring territories.

During the ninth century, the kingdom began to decline. The rulers of the Idrisid Dynasty were eventually crushed by the Umayyad caliphs of Cordoba, Spain, and the Fatimids of Cairo, Egypt. The last Idrisid ruler was killed while a prisoner of the Umayyad in 985.

Two hundred years later, a religious reformation generated another empire and metropolis. The Almoravids began as a confederacy of religious warriors, Berbers who founded a brotherhood in secluded fortified retreats. In 1062 the Almoravids founded the city of Marrakesh, and by the end of the century they had expanded their empire east to Algiers, south to the Senagal River, and north to the Ebro River in Spain.

During the next century, however, the Almoravid's political power waned. In 1146 the Almohads massacred the inhabitants of Marrakesh and seized control of the region. Under the Almohads, Morocco became the center of an empire that embraced modern-day Algeria, Tunisia, Libya, and large areas of Spain and Portugal.

The Almohad Empire began to disintegrate after the *Battle of Las Navas de Tolosa* on July 16, 1212. Also called *Battle of Al-'ugals,* this confrontation was a major step in the Christian reconquest of Spain, in which the Almohads were severely defeated by the combined armies of Léon, Castile, Aragon, Navarre, and Portugal. By mid-century the power and prestige that had once belonged to Morocco was gone. A period of disorder and almost incessant civil war between Berbers and Arabs followed. Rulers of various dynasties reigned briefly and ineffectually over parts of the country.

Morocco experienced a revival of power under the Saadians. Known as the first *Sharifian Dynasty,* 1554–1660, the reign of

Jewish Men, Tripoli, 1925   *It is estimated that 1.5 million Jews have lived for several centuries in North Africa and the Middle East—and their ancestors have never lived in Europe. These people are referred to as Oriental Jews. In the Arab lands of Morocco, Algeria, Tunisia, Libya, Egypt, Yemen, Jordon, Lebanon, Iraq, and Syria, Oriental Jews spoke Arabic as their native language.*

In 1920 it was near estimated that the population of Tripoli stood at 50,000—10,000 Jews; 35,000 Arabs and Berbers; and 5,000 Europeans.

Following the establishment of Israel in 1948, practically all of the Yemenite, Iraq, and Libyan Jews migrated to Israel.

Ahmed I al-marr-sur is regarded as the golden age of Morocco. The country benefited enormously from the influx of nearly a million Moors and Jews who were expelled from Spain after 1492. It experienced a period of unity and prosperity in which arts and culture flourished.

The Saadians were succeeded by the second Sharifian dynasty, which has ruled Morocco since 1660. This dynasty reached its peak with the reign of Ismail al-Hassani (1672–1727). Al-Hassani's reign was followed by a long period of disorder punctuated by brief periods of relative peace and prosperity.

In 1578, during the reign of the first Sharifian dynasty, Morocco inflicted a devastating defeat on the Portuguese, and by

Morocco

Jewish Women, Tripoli, 1925

the seventeenth century Moroccans had regained control of most of their coastal cities. In the eighteenth and nineteenth centuries pirates from Morocco attacked European ships along the Barbary Coast of the Mediterranean Sea. Because of the continued harassment suffered at the hands of the Barbary pirates, and because Morocco shared control of the Strait of Gibraltar with Spain, the country came to the increasing attention of the European maritime powers, especially Spain, Britain, and France. Spain invaded Morocco in 1859–1860 and acquired Tetovan.

In April 1904, in return for France's support in Egypt, the British government recognized Morocco as a French sphere of

North Africa

Jewish Quarter, Tripoli, 1925

Morocco

Fortress of the Kaid of Shaonia, 1896 *Sir Arthur Nicholson travelled inland to the High Atlas Mountains from the port of Mogador (now Essaouira). An experienced British diplomat, he had instructions to create a strong Moroccan government under the Kaid of Shaonia with British protection. His mission failed. In 1904 Great Britian allowed France a free hand in Morocco in return for a French guarantee not to interfere with British domination of Egypt and the Sudan.*

interest. Later that year France and Spain divided Morocco into zones of influence, with Spain receiving the smaller part of Morocco and the region south of Morocco, which would become Spanish Sahara. Imperial Germany soon disputed these arrangements and, as a result of this dispute, a conference of the major powers of the world, including the United States, was held in Algeciras, Spain, in January 1906, to arbitrate the matter. The resultant *Act of Algeciras* guaranteed equality of economic rights for every nation in Morocco. Of course, these decisions were made at the expense of the people of Morocco and were the cause of continued civil unrest in the country. For several years Morocco teetered on the brink of civil war.

In July 1911 the Germans sent a gunboat to the Moroccan port city of Agadir in a thinly disguised move to intensify the unrest brewing in the country. This incident provoked the French and brought Europe to the brink of war, but in later negotiations Germany agreed to a French protectorate over Morocco in return for French territorial concessions elsewhere in Africa. Having few alternatives, the sultan of Morocco submitted to the will of the European invaders. His acquiescence was symbolized by the signing of the *Treaty of Fez* on March 30, 1912.

The treaty gave France control over Morocco's foreign relations, police powers, and finances. Although it stripped the rulers of Morocco, present and future, of most of their powers, it did grant them the right to veto protectorate legislation, a right that would become critical during the nationalist struggle to regain independence. The treaty left about a tenth of the country under Spanish control, including parts of present-day Western Sahara, and granted Tangier special status as an international zone.

Abd el-Krim led the first major anticolonial revolt in the Spanish-held territories, where he founded the *Rif Republic* in 1921. By 1924 he had driven the Spanish forces from most of their Moroccan territory. He then turned his attention to the French. As a result, France and Spain agreed in 1925 to cooperate in their efforts to defeat Abd el-Krim. More than 200,000 troops under the French marshal Henri Phillippe Pétain were used in the campaign against the Moroccan rebels. Although Abd el-Krim was defeated by 1926, it took another eight years to suppress the rebellion.

The nationalist struggle for liberation was diverted during World War II (1939–1945). In November 1942 American troops landed and occupied Morocco. During the rest of the war, the country was a major Allied supply depot. Casablanca was the site for the meeting of the heads of government of the Allies in 1943.

In 1944 Moroccan nationalists formed the *Hizb al-Istiqlal,* or Independence Party, which drafted a *Manifesto of Independence.* Although it was supported by Sultan Mohammad V and

Morocco

the majority of Arabs, the French responded by arresting the leaders on accusations of Nazi collaboration.

Opposition to French rule increased during the 1950s, especially after a trade union protest in Casablanca led to hundreds of arrests in 1952. By this time, Istiglal had more than 80,000 members and several hundred thousand sympathizers, including many from Berber communities. The French government, meanwhile, was increasingly preoccupied with civil unrest in Algeria and other parts of its colonial empire.

The French government recognized Moroccan Independence on March 2, 1956. In April the Spanish government recognized in principle the independence of Spanish Morocco and the unity of the sultanate, although it retained certain cities and territories. Tangier was incorporated into Morocco in October 1956. Ifni, a former Spanish territory, was returned to Morocco in January 1969.

Sultan Mohammad V assumed the title of king in August 1957. At his death in 1961, the throne passed to his son, Hassan II. A royal charter was instituted by Hassan in which a constitutional monarchy was established by referendum of a constitution in December 1962. The nation's first general elections were held in 1963. In June 1965, however, the king temporarily suspended parliament and assumed full executive and legislative power, serving as his own prime minister for two years. Hassan gave strong support to the Arab cause in the June 1967 war against Israel. He also supported the Palestinian effort to regain their homeland. Even so, he was accused of being too moderate, and two attempts were made on his life in 1971 and 1972.

During 1974 and 1975 Morocco exerted pressure on Spain to relinquish Spanish Sahara. When the Spanish left in 1976, they surrendered the northern two-thirds of the colony to Morocco, while Mauritania received the southern third. This partitioning of the phosphate-rich territory was disputed by the *Polisario Front,* a Saharan nationalist guerrilla movement that proclaimed Western Sahara an independent nation, called the *Sahrawi Arab Democratic Republic (SADR).* The Polisario Front (*Frente Popular para la Liberacion de Saquia El-Hamra y Rio de Oro*) was founded May 10, 1973, at a secret meeting on the border of Western Sahara and Mauritania. It has struggled for nearly thirty years to gain national independence for Western Sahara. Polisario chose to focus first on Mauritania. In 1979 it forced Mauritania to relinquish its claims in Western Sahara and to recognize the SADR.

Polisario's campaign against Morocco continues today but has been less successful. Although it managed to inflict heavy losses on Moroccan forces during the 1980s, Moroccan military engineers limited Polisario military incursions by encompassing Western Saharan in an earthen wall protected by motion detectors, explosives, and soldiers. However, both Polisario and Morocco faced increasing financial difficulties in the 1990s, and in 1991 the United Nations brokered a cease-fire agreement, pending a referendum. Disagreements between Polisario and the Moroccan government over the eligibility of voters have repeatedly postponed the referendum.

The conflict between Morocco and Polisario took a critical turn on July 23, 1999, when King Hassan II died of a heart attack and his eldest son, Muhammad ibn al-Hassan, or Mohammad VI, succeeded him to the throne. Although the new king pledged to continue his father's policies, the tensions between Morocco and the Polisario Front have lessened somewhat in the years following the older king's death. In fact, in December 2000 the Polisario released 200 Moroccan soldiers that had been captured twenty-five years earlier, at the start of the conflict over Western Sahara. Nevertheless, as of April 2002, over 1,000 Moroccan citizens are still being held prisoner by the Polisario Front—which means that the dispute is not settled.

In spite of these internal conflicts, the kingdom of Morocco, *Al Mamlakah al Maghribiyah,* is a nation known by its cities. Rabat, the capital; Casablanca, the country's largest city and central seaport; Marrakesh and Fez, both important trade centers; and Tangier, a seaport on a bay of the Strait of Gibraltar, have all played crucial roles in Morocco's dynastic history. Each one in turn has served as political, economic, and cultural capitals of the kingdoms that compose what is now Morocco.

Morocco

For centuries these cities also have provided vital centers for the commerce in goods and ideas that came from the Islamic world and Christian Europe as well as sub-Saharan Africa.

In the nineteenth century, as European investment poured in, Morocco's cities held perhaps too much appeal. The gravitation of people and resources toward Morocco's cities, especially on the coast, zapped the vitality of the rural agricultural economy on which the urban prosperity had depended. After becoming perilously indebted to Europe, Morocco fell under French control in the early twentieth century. During this time European and American artists, writers, and wanderers were drawn by a vision of the *exotic* beauty of Morocco. Today Morocco's economy still depends largely on agriculture, but its cities remain dynamic commercial centers as well as destinations for migrants seeking a better life and travelers in search of legends.

Constantine, c. 1880 *This is a photograph on Rue Arabe in the old quarter of Constantine, a walled city in northeast Algeria near the border with Tunisia.*

3

TUNISIA

Eight centuries before the birth of Christ, Tunisia became the stage for the extraordinary drama of Queen Elissa, founder of the ancient city of Carthage. Elissa is known to Western readers through the Roman poet Virgil and his epic poem the *Aeneid,* in which she is called Queen Dido.

Dido was the daughter of Belus, king of Tyre, the most important city of ancient Phoenicia. Belus was located at the site of present-day Sur in southern Lebanon. Herodotus, the Greek historian, records a tradition that traced the settlement of Tyre back to the twenty-eighth century B.C. The town is frequently mentioned in the Bible, Old and New Testaments, as having close ties with Israel. Haram, king of Tyre (reigning 961–936), furnished building materials for Solomon's Temple in Jerusalem, and the notorious Jezebel, wife of King Ahab, was the daughter of Ethbaal, king of Tyre and Sidon.

When Dido was old enough to be married, arrangements were made for her to wed her uncle Archabas, the high priest. Their happiness was shattered shortly after their wedding vows by Dido's brother Pygmalion, who not only usurped his father's throne but also murdered Archabas in the process. In the seventh year of her brother's reign, Dido decided to flee Tyre for the West. She was accompanied by several members of

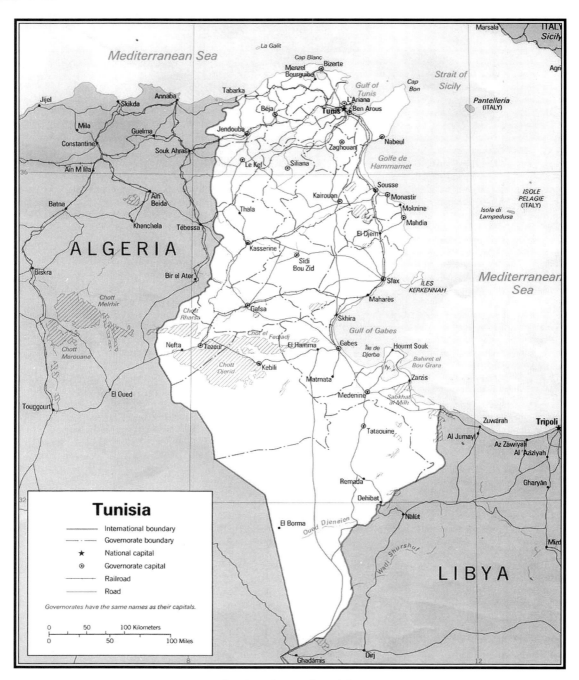

Modern Map of Tunisia

Tunisia

The Ruins of Carthage, 1983 *Carthage was one of the great cities of antiquity, dating back to the eighth century B.C. A wealthy trading center, it stood on a peninsula near the present city of Tunis. Legend tells us that Dido, daughter of the king of Tyre, a Phoenician city, founded Carthage. The story of her tragic love for Aeneas, a Trojan prince whose family founded Rome hundreds of years later, is the subject of Virgil's epic poem, the* Aeneid.

Carthage was the first city-state to control an empire. Much of western North Africa, southern Spain, Sardinia, Corsica, and the western half of Sicily came under Carthage's rule. From the middle of the third century B.C. to the middle of the second century B.C., Carthage was involved in a series of wars with Rome known as the Punic Wars. They ended with the total defeat of Carthage in 146 B.C. The city was plundered and burned—the Romans forbade all human habitation on the site.

Subsequently, the Romans built a new city near the ancient site. New Carthage soon ranked with Alexandria and Anticoch as one of the great cities of the Near East. St. Augustine was one of its famous inhabitants. The final destruction of New Carthage by the Arabs came in A.D. 698.

The building in the rear of the photograph is of the Cathedral of St. Louis. In 1270 Louis IX of France (St. Louis) led the Eighth Crusade. He died en route at Tunis.

the royal family who had fallen out of favor with Pygmalion. Her aristocratic party stopped at Cyprus, where they were joined by several other refugees, including the priest of the Temple of Juno and his family. They also abducted eighty maidens to take along on their journey. From Cyprus they sailed directly to the future site of Carthage—three low hills and two small lagoons sitting in a broad isthmus flanked by two large salt lakes. Eventually, the Tyrian princess and her followers encountered the indigenous inhabitants of the region, the Berber people of North Africa and their ruler, Iarbas. According to legend, Iarbas and Dido struck a deal that gave her as much land as could be covered with an ox hide. Being a shrewd negotiator, Dido had a hide cut into the thinnest possible strips, with which she marked the boundaries of the new city. Impressed by the young woman's ingenuity, the Berber king kept his end of the bargain and gave her land enough to build her city, and the bargain they made that day became forever interwoven into the history of Tunisia. The high citadel at the center of Carthage continues to this day to bear the name *Byrsa,* Greek for "ox hide."

Unfortunately for Dido, the king's fascination with the crafty princess did not end there. Over time he began to press for her hand in marriage. Concerned that marriage to the Berber king would require her to move inland and abandon her people and the city she had founded, the princess refused his advances. Afraid of endangering her people by continuing to refuse the importunate king, she built a funeral pyre of sandalwood outside her palace and leapt into the flames.

Virgil tells a drastically different version of this story. According to his version, Aeneas, the Trojan prince and the future founder of Rome, was shipwrecked at Carthage after escaping from Troy at the end of the Trojan War. His aging father, his son, and other survivors of the war accompanied Aeneas. Dido, who had pledged herself to celibacy after the murder of her husband, received the shipwrecked strangers and, over time, fell in love with Aeneas. The two began to live together as husband and wife. When it became clear that Aeneas intended to make Carthage his home, Jupiter warned him that he must leave Dido and fulfill his destiny of becoming the founder of Rome. In despair at his departure, Dido killed herself on a funeral pyre. Later, during his journeys, Aeneas encountered the ghost of Dido in Hades, but she refused to speak to him.

Both versions of the story end in tragedy and death for the first queen and founder of the ancient city of Carthage. However little of either story might be historically accurate, the fact remains that the founding of Carthage, *Oart Hadasht,* or New City, in 814 B.C. by Elissa (Dido), remains a decisive historical event and a potent legend in world literature. Chapter IV of Virgil's epic masterpiece has inspired an opera by Thomas Purcell, paintings by Joseph Turner, and literature by African-American writers W.E.B. Dubois and Charles Chestnut. Even the early Christian bishop St. Augustine was not immune to the power of the tragic story of Dido. He remarks in the *Confessions,* "*. . . because I wept for Dido, who killed herself for love.*"

Today Carthage is a wealthy suburb of Tunis, the capital city of Tunisia, but at its zenith, Carthage was the center of a commercial empire that dominated trade in the western Mediterranean. Its reign of supremacy lasted three centuries, from 500 to 200 B.C. The ancient Carthaginians probably emerged from the Arabian Peninsula about 5,000 years ago. Called *Sidonians* in the Old Testament and *Phoenicians* by the Greek poet Homer, they were a Semitic people related to the Canaanites and Philistines of ancient Palestine. After subduing the indigenous peoples of Syria and Palestine, they established a maritime empire at the eastern end of the Mediterranean. The Phoenicians worshiped a paramount god, Baal, and other divinities. They also invented the alphabet. It is believed that the Phoenicians circumnavigated Africa in the sixth century B.C. and sailed as far north as Britain, where many Carthaginian coins have been found.

Carthage, the city on the bay, exemplified everything the Phoenicians held dear. Well situated for a maritime nation, it offered anchorage for many ships. Its central Mediterranean location was as close to Europe, at least to Sicily, as any other place on the African coast east of Morocco. The distance from Carthage to its colonies in western Sicily was no more than a day's journey for the swift Carthaginian vessels. The narrow opening between the eastern and western Mediterranean could be patrolled and if necessary closed by a line of warships. The only way around was to pass through the perilous Straits of Messina, which meant sailing between the dreaded *Scylla* and *Charybdis,* two sea monsters in Greek mythology dwelling on the opposite sides of a narrow strait, personifications of the danger of navigation near the rocks and eddies there.

Scylla had twelve feet and six heads on long, snaky necks. Each head had a triple row of sharklike teeth, with which she devoured any prey that came within reach. Across the strait, a bowshot away, was a large fig tree under which Charybdis, the whirlpool, lurked, drinking and belching forth the waters of the Mediterranean three times a day, engulfing anything that came near. According to the legend, when Odysseus was sailing home to Ithaca he passed between them. His ship was able to avoid Charybdis, but Scylla seized six of his men and devoured them.

The high point of the Carthaginian hegemony occurred around 400 B.C. It was during this time that Carthage founded settlements that the Greeks called *emporia* along the entire coast, from the Gulf of Sidra in present-day Libya, through present-day Tunisia and Algeria, to the Atlantic coast of Morocco, and on all the islands of the western Mediterranean including the Balearics. Invading ships were sunk when captured, and a large army of Libyan and Nubian mercenaries countered invasions or put down revolts. But always, the Carthaginian trade flourished, not only by sea but also across the Sahara, along routes that opened much of sub-Saharan Africa to trade and commerce, especially in gold and precious jewels, for which the Carthaginians exchanged cloth and manufactured items. At the height of its glory the city had as many as 500,000 inhabitants, and was reported to be the wealthiest city in the Mediterranean world.

The beginning of Carthage's decline came in 264 B.C., with the onset of the *First Punic War,* and its total defeat and devastation came with the last Punic war.

Tunisia

The First Punic War was the outcome of intense political and economic rivalry between Carthage and Rome. Both countries wanted to establish control over the strategic islands of Corsica and Sicily. In 264 B.C. the Carthaginians intervened in a dispute between the two principal cities on the Sicilian West Coast, Messana (now Messina) and Syracuse, and so established a presence on the island. Rome, not wanting to lose face, attacked Messana and forced the Carthaginians to retreat. Rome's attack provoked a declaration of war from the Carthaginians. The war lasted twenty-five years and proved a disaster for Carthage. Not only was Sicily lost, but also Sardinia, Corsica, and Malta, along with the monopoly of trade west of Italy.

Hamicar Barca, a distinguished Carthaginian general of the First Punic War, devoted the remainder of his life to building up Carthaginian power in Spain to compensate for the loss of Sicily. His son Hannibal, known as one of the greatest military leaders in history, became commander of the Carthaginian forces in 221 B.C. and in 219 B.C. he captured Sayuntum, a Spanish city allied with Rome. This attack provoked the *Second Punic War* (218–201 B.C.) Because Rome controlled the sea, Hannibal led his army of 40,000 infantry, 10,000 cavalry, and fifty elephants through Spain and Gaul, and then across the Alps to attack the Romans in Italy before they could complete preparations for war. His passage through the rugged mountains is one of the great feats in military history. By 216 B.C., he had won two major victories, one at Lake Trasimeno and the other at Cannae. The latter is one of the most famous battles in European history.

Meanwhile, although Hannibal and his exhausted troops had proven victorious over the Romans on their home turf, a Roman general by the name of Publius Cornelius Scipio Africanus, known as Scipio Africanus the Elder, had totally defeated the Carthaginians in Spain, and in 204 B.C. he landed his army in North Africa. The Carthaginians recalled Hannibal to Africa. Leading an army of mostly untrained recruits, he was decisively defeated at the *Battle of Zama* in 202 B.C. This battle marked the end of Carthage as a great power and the close of the Second

Punic War. The Carthaginians were forced to surrender Spain and the islands of the Mediterranean still in their possession, disband their navy, and pay tribute to Rome. Hannibal lived another twenty years and continued to wage war against the Roman Empire until he was finally trapped in Bithynia, a small village near the Black Sea, where he ended his life by drinking poison. The year was 183 B.C.

The first and second Punic wars, 264–241 B.C. and 218–201 B.C, had effectively deprived Carthage of its political and economic power. Nevertheless, the city slowly began to prosper again. By 150 B.C. it was once more rich and influential. This resurrection of Carthage from the ashes of its terrible defeat incited the wrath of the Romans who still remembered the humiliating defeats Rome had suffered at the hands of the Carthaginians. In particular, the Roman statesman Cato the Elder became obsessed with the idea that Carthage was a menace to Rome—so much so that until his death in 149 B.C., he concluded every speech before the Roman Senate, regardless of the subject, with the words *Delenda est Carthago,* "Carthage must be destroyed." In the year of his death, largely due to his influence, war between Carthage and Rome, the *Third Punic War,* began, resulting three years later in the complete destruction of Carthage. The city was besieged, captured, and plundered, its walls demolished, its houses and public buildings burned to the ground. To ensure that it would never again rise from the ashes of its defeat, its smoking ruins were sown with salt. Of a city population that had exceeded 250,000, only 50,000 survived the final surrender. The survivors were sold into slavery. Twenty years later a Roman colony was constructed on the site, and in due course New Carthage became the capital of the Roman province of Africa, the home of modern Tunisia.

Tunisia has a long and distinct history as a politically and culturally unified country despite its subjection to a variety of rulers and the influences of contrasting civilizations over a period of nearly 3,000 years. Modern Tunisia derives its name from Tunis, which was originally founded by the ancient

Tunisia

Libyans but became part of the Carthaginian Empire in the nineth century and since the thirteenth century has served as the country's capital and principal city. To the early Romans and later the Arabs, this relatively small region was known as Africa or *Ifriqya,* a name that was eventually extended to the entire continent.

History-conscious Tunisians point to Carthage and Kairouam as proof of their continuous history and development as a people and a nation. Carthage built an empire that dominated North Africa and the western Mediterranean until it fell victim to the phalanxes of invading Roman armies. Destroyed by its conquerors and then resurrected from the ashes as the administrative center of *Roman Africa,* the city of Carthage became in time the spiritual center of Latin Christianity in North Africa. Kairouan, or Al Qayrawan, as mentioned earlier, was founded in the seventh century A.D. Called the "City of One Hundred Mosques," it is one of the holy cities of the Muslim world, a fountain from which Arab and Islamic culture has flowed across North Africa.

The architecture of Al Qayrawan, with its Moorish and Saracenic inscriptions, dates from the ninth century. Even older Kufic inscriptions and Roman ruins are also found there. Turreted walls and gates give the city the aspect of a medieval Arab fortress. Founded by an Arab emir in 670, it became the capital of the Ifriqya province of the caliphate in the eighth century and of the *Aghlabid Dynasty* in the ninth century. In the eleventh century, Al Qayrawan was capital of the *Zirid Dynasty,* and in the fifteenth it was an administrative center under the Almoravids. It was occupied by the French in 1881.

Tunisia was once part of the great medieval Berber empire, the most illustrious city of which was Numidia, during the reign of Masinissa (240 B.C.–148 A.D.). From the sixth century B.C. the Carthaginians expanded into the interior, and by the third century B.C., they had reached as far as Theveste, or Tebessa, and occupied points along the coast of Numidia. Numidians were frequently found in the Carthaginian army. In fact, during the Second Punic War, Masinissa fought as Hannibal's ally, although

Great Mosque, Al Qayrawan, 1893 *Al Qayrawan, one of the holy cities of Islam, is located in north-central Tunisia about 95 miles south of Tunis. By the eleventh century, the city had become one of Islam's great administrative, religious, and intellectual centers. The present Great Mosque, originally built in the seventh century, is the fifth constructed on the site.*

Tunisia

he switched his allegiance in 206 and was given further territory, extending his empire as far as the Moulouya River.

After Masinissa's death in 148 B.C., the Romans divided his kingdoms among several less powerful rulers. Caesar formed a new province, *Africa Nova* ("New Africa") from Numidian territory, and Augustus united Africa Nova with *Africa Vetus* ("Old Africa") from the province surrounding Carthage. But a separate province of Numidia was formally created by Septimius Severus.

Christianity spread rapidly in the third century A.D., but in the fourth century Numidia became the center of the aforementioned Donatist movement. This schismatic Christian group was particularly strong among the Numidian peasantry, to whom it appealed as a focus of protest against worsening social conditions. After the Vandal conquest of 429 B.C., Roman civilization had declined rapidly in Numidia, and the indigenous populations revived to outlive, in many respects, even the Arab-Islamic conquest of the eighth century and to persist until modern times.

For centuries, from the late seventh century to the early sixteenth, North Africa was part of the Islamic empire of the Maghrib, in which Arab conquerors replaced Roman-Christian culture with the Islamic way of life as a succession of dynasties wielded power. Most notable among them were the Aghlabids (800–909), the Fatimids (909–973), and the Zirids (tenth century). Then, in the latter part of the twelfth century the *Normans*, led by Roger II, the first king of Sicily (1130–1154), briefly occupied a number of important coastal regions. The Normans were members of a Scandinavian people, *Vikings*, or *Norsemen*, who settled in northern France, creating the *Frankish* kingdom in the tenth century. They founded the duchy of Normandy and sent out expeditions of conquest and colonization to southern Italy, Sicily, England, Wales, Scotland, and Ireland. The most important of these endeavors, known as the *Norman Conquest*, was the invasion of England in 1066 (Battle of Hastings) by William, Duke of Normandy, who subsequently became king of England.

Early in the eleventh century, Norman adventurers also began a somewhat more prolonged and haphazard migration to southern Italy and Sicily, where they served the local nobility as mercenaries fighting Arabs and the Byzantines. As more Normans arrived, they carved out small principalities for themselves. Some of these were on the coast of North Africa. The Arabs recovered the region later in the century, and the *Almohad dynasty* of the twelfth century and the *Hafsid dynasty* of the thirteenth succeeded to power.

Arab political, economic, and cultural supremacy came to an end in the early fifteenth century. During the period of Arab-Islamic domination, the region came to be known as *Tunis,* or *Tunisia,* after its chief city. Even from these early days, Tunisia has often been described as an oasis in the desert. This image of a cool, refreshing sanctuary in the heart of a hot, burning climate refers both to the country's natural beauty, which has attracted wanderers and seekers for centuries, and its political and social climate. In regard to its breathtaking natural beauty, Tunisia is bounded on the north and east by the Mediterranean coast, the same coast that first attracted Dido and her companions. It is indented by many harbors and inlets, notably the Gulf of Tunis, Hammemet, and Oabis. The Gulf of Gabes contains the islands of Jarqah, or Djeba, and Qarqanah, or Kerkennah. In general, a mild Mediterranean climate prevails in Northern Tunisia. The fertile, well-watered regions of the north are characterized by flourishing vineyards and dense forests of cork, oak, pine, and juniper trees. In the arid regions of the extreme south, date palms flourish in oases.

Because of its enchanting environment and calm political and social climate, Tunisia has promoted itself as a secular haven from the troubles of the rest of the Arab World. Indeed, the country has been on the vanguard of Western-inspired reform since the ninteenth century. Yet populist support continues to grow within the nation for an Islamic party, and Tunisia has never, like Morocco, really been isolated from its North African neighbors, neighbors with whom it shares the religion of Islam and the legacy of Phoenician, Roman, Arab, and European conquest.

Tunisia

Well, Sahara Oasis, 1896   *An Oasis is a fertile area of land in a desert. Underground water sources account for most oases. In all Saharan oases, the date palm is the main source of food.*

In previous centuries piracy lured adventurers from around the Mediterranean to the Maghribi coastal cities and islands. Among them were two brothers Aruj and Khair al Din, the latter known to Europeans as Barbarossa or Redbeard. The brothers reached Tunisia in 1504 and sailed from Jerba Island under Hafsid patronage. Muslims from the Greek island of Lesbos came as well. During the thirteenth century the Hafsid monarchy, Berber descendants of the *al-Muwahhid Dynasty,* had risen to power. They shifted their capital from the interior to coastal Tunis, near the ruins of Carthage, signifying an increased emphasis on maritime trade in a region that would soon and ever after become known as Tunisia. In 1510 the Barbarossa brothers, as they became known in the West, were invited by Algiers to help defend it against the Spanish. Instead, they seized Algiers and used it as a base of operations not only for piracy but also for conquest in the interior. Khair al Din eventually recognized the authority of the Ottoman Empire over the territory that he controlled and was in turn appointed the Ottoman's regent in the area, bearing the title of *beylerbey,* or "commander in chief." He was forced to abandon Algiers temporarily from 1519–1529 to the Hafsids, who resisted Ottoman control of the Maghrib. But with the assistance of Turkish troops, Khair al Din was able to consolidate his position in the central Maghrib and in 1534 mounted a successful seaborne assault against Tunis.

The Hafsid sultan, Hassan, took refuge in Spain, where he sought the aid of the Hapsburg king-emperor Charles V, to restore him to his throne. Spanish troops and ships recaptured Tunis in 1535 and reinstalled Hassan. Protected by a large Spanish garrison at La Goulette, the harbor of Tunis, the Hafsids became the Muslim ally of Catholic Spain in its struggle with the Turks for supremacy in the Mediterranean, making Tunisia the stage for repeated conflict between the two great powers.

In 1574 armies of the Ottoman Turkish Empire defeated the Spanish and assumed hegemony over Tunisia. Under Ottoman rule, Tunisia enjoyed a period of relative stability from 1574 to 1881. Imperial rule was carried out through local administrators, who were known as *deys* of Tunis until 1705 and as *beys* thereafter. The first bey, al-Husayn ibn Ali, reigned from 1705 to 1740, and founded the *Husaynid dynasty,* which secured for Tunisia a limited degree of autonomy and a large measure of prosperity.

Unfortunately, this stability and prosperity was not meant to last. In the mid-eighteenth century, a series of plagues swept through North Africa, killing thousands of Tunisians in its wake. The death toll, combined with a simultaneous prolonged drought, sent the agricultural economy into a crisis. During

Tunisia

1803 two other factors conspired to drive the Maghrib deeper into catastrophe. One was the inevitable, irrevocable decline of the Ottoman Empire and the other was the ever encroaching tide of European colonialism. France annexed Algeria in 1834.

Hoping to maintain autonomy in the face of European imperialism, the Tunisian ruler Ahmad Bey set out to strengthen the Tunisian state through modernization. He modeled his government after European bureaucracies, conscripted peasants for his greatly expanded army and navy, and imposed heavy taxes to pay for both. Under Ahmad Bey, Tunisia became the first country in the Islamic World to abolish slavery. However, no matter how progressive these measures were, they were also costly and plunged Tunisia deeper into debt. Although Ahmad Bey hoped that loans from European banks would help alleviate the financial crisis, the Tunisian economy stagnated as it fell deeper and deeper into crisis, which in turn led to civil unrest. The final blow came in 1869, when the country declared bankruptcy, and an international financial commission consisting of France, Great Britain, and Italy was formed to oversee the monarch's finances.

Ironically, the same countries appointed to remedy Tunisia's financial woes, France, Italy, and Great Britain, were also the country's chief creditors, and all of them had imperialistic ambitions in North Africa. The French, to get a jump on its competition, invaded Tunisia from Algeria in 1881, using the pretext that some Tunisians had crossed the border into French-controlled Algeria. On May 12, 1881, the reigning bey of Tunisia signed the *Treaty of Kasser Said,* known also as the *Bardo Treaty,* which acknowledged Tunisia to be a French protectorate. The two countries signed the supplemental *Convention of Marsa* in 1883.

During the early 1900s opposition to French colonization gave birth to several militant independence movements known collectively as the *Young Tunisians.* For several decades French authorities successfully suppressed these fledgling patriotic movements. In 1920, however, various nationalist groups united and formed the *Destour,* or "Constitutional Party," which advocated extensive democratic reforms. The Destour movement was disbanded in 1925, but it was revived during the economic depression of the 1930s. In 1934, the so-called *Neo-Destour*, or "New Constitutional Party," was organized by the Tunisian patriot and statesman Habib Bourguiba, the future first president of Tunisia, 1957–1987. In contrast to the more moderate Destour Party, which looked for support only in Tunisia, the Neo-Destour Party sought and received aid from extreme leftist and nationalist groups in France, Morocco, and Algeria. By 1937, the party had 28,000 activists and nearly 50,000 supporters operating in 400 village branches.

The Neo-Destour was the vanguard of future decolonization struggles, using acts of civil disobedience, including a general strike in solidarity with nationalist's movements in other North African countries. In 1938 the Neo-Destour party was outlawed, and Bourguiba and other leaders were arrested and deported to France. Four years later, when the Germans occupied France and Tunisia during World War II, Bourguiba was released. Despite his refusal to support the Axis powers, the coalition of countries that opposed the Allied powers in World War II, he was allowed to return to Tunis, where he took a leading role in the nationalist struggle emerging in German-occupied Tunisia.

In 1945 the French forced Bourguiba to flee from Tunisia and seek refuge in Egypt. In the following year France, still under threat of rebellion from nationalist groups, granted Tunisia status as a "semiautonomous" associated state of the French Union. Further steps toward autonomy came in August 1947, when the French resident general formed a ministry composed primarily of Tunisians. The French, however, retained the majority of political power.

In September 1949 Bourguiba returned from exile and resumed his campaign for Tunisian independence. France, responding to the increasing militancy of nationalist feelings of the Tunisian people, appointed more Tunisians to government posts and positions in the civil service. The following year, 1952, Tunisians ministers attempted to voice their grievances against the French government before the United Nations, but they were prevented from doing so by a ruling that the dispute involved a domestic rather than an international issue and therefore did not fall within the jurisdiction of the United Nations. In the meantime, revolts and political demonstrations occurred almost on a daily basis, especially in Northern Tunisia, rendering the French authority increasing untenable. The rebellion continued almost unchallenged throughout the first six months of 1954, during which time the French made repeated offers of limited concessions. Their offers went unheard.

Anti-French revolts became increasingly violent in July 1954. The French government felt the need to take desperate measures, and on July 31, 1954, the French premier, Pierre Mendes-France, arrived in Tunisia on a mission of conciliation. He promised the Tunisian people full internal autonomy under a government composed of Tunisians. This promise was enough to bring an immediate halt to the rebellion that had threatened to completely engulf the country. In the calm that followed, the two countries engaged in lengthy negotiations. As a result, on June 3, 1955, after almost a year of continuous discussion, the Tunisian premier, Tahar ben Ammar, and the new French premier, Edgar Faure, signed a series of conventions and protocols that greatly increased Tunisian independence. The stumbling block came when France insisted on retaining control of Tunisian foreign policy and defense. Nevertheless, on September 17, 1955, the first all-Tunisian government in seventy-four years took power in Tunis.

Almost immediately, many nationalist militants opposed the new regime and clamored for an even greater measure of independence. Confronted with this dilemma, the French made further concessions, symbolized in the historic protocol signed March 20, 1956. The agreement annulled the Bardo Treaty of 1881 and recognized Tunisia as a completely sovereign, constitutional monarchy governed by the bey of Tunis. The first national legislative elections in Tunisian history, which took place on March 25, 1956, resulted in a decisive victory for the Neo-Destour Party. On April 8, Habib ibn Ali Bourguiba was elected president of the *Tunisian National Assembly*. Three days later, he was named premier. The assembly adopted a constitution transferring to the Tunisian people the legislative powers previously exercised by the bey. On November 12, 1956, Tunisia was admitted to the United Nations. The following year, on March 5, 1957, the political strength of the Neo-Destour Party was demonstrated once again when it received 90 percent of the votes in various municipal elections. Noteworthy was the fact that women voted for the first time.

During the first years of the new republic, the war for independence in neighboring Algeria strained relations between Tunisia and France. The French government accused Tunisia of supporting the Algerian nationalist struggle for liberation, and in 1958 bombed the Tunisian village of Saqigat Sidi Yusuf. A United Nations–mandated cease-fire eventually eased military tensions, but France withdrew all financial support. In turn, Tunisia forged alliances with other Arab countries, especially Saudi Arabia.

In the 1960s and 1970s Tunisia pursued peaceful economic development, particularly of its petroleum resources. Relations with France improved, but Bourguiba expressed his distrust of the United States and Soviet intentions in the Middle East. In 1982 Tunisia gave refuge to the Palestine Liberation Organization (PLO) leader Yasir Arafat and several hundred of his followers who had been forced out of Lebanon.

In November 1987 Prime Minister Zine al-Abidine Ben Ali assumed the presidency when President Bourguiba, after thirty years in office, was declared incompetent due to his age. Thirteen years later, on Thursday, April 6, 2000, Habib ibn Ali Bourguiba died in solitude five months before his ninety-seventh birthday. On the gates of the family mausoleum where he was buried, the following words were written, "The supreme combatant, builder of the new Tunisia, women's liberator."

Although several parties contested the elections of April 1989, Ben Ali's *Democratic Constitutional Rally Party* won all 141 seats in parliament, and Ben Ali was elected to the presidency unopposed. In 1994 and again in 1999 Ben Ali was reelected to the presidency. In a major address to the nation on

Tunisia

the occasion of his accession to office, President Zine al-Abidine Ben Ali announced a major constitutional reform to pave the way for "the republic of the future." He proclaimed, "Our commitment to republican values is firm and unshakable. We believe in the sovereignty of the people, the primacy of the constitution and the inviolability of the institutions. We also believe in the values of liberty, democracy, justice, pluralism and human rights."

Postcard Picture for Tourist Shops, 1896 *The opening of the Suez Canal in 1869, together with the advent of the steamship, continued European industrialization, and French colonization in North Africa made the Mediterranean again one of the world's busiest sea lanes. By the 1870s, Thomas Cook & Son were conducting organized tours through North Africa. In winter, affluent Europeans came to the North African coast seeking warmer weather. This is a studio card photograph, a posed picture sold in tourist and gift shops.*

4

LIBYA

The year 1997 was a decisive one in the life of Nelson Mandela, the first black president of South Africa. It was the year he made the decision not to run for reelection, even though his popularity almost guaranteed him another victory. Nevertheless, he decided to step down, bringing an end to an era that had begun almost a century before. On December 20, 1997, during a speech delivered at the closing session of the Fiftieth National Conference of the African National Congress, Mandela announced, "The time has come to hand over the baton."

It was a decision that shocked and dismayed his supporters and critics, and stunned members of the audience listening to his resignation speech. Three years earlier, on April 27, 1994, South Africa held its first "all-race" elections, with black South Africans voting for the first time in their lives, and Nelson Mandela was their candidate of choice. He was inaugurated as president on May 10, 1994. As president, Mandela established the *Truth and Reconciliation Commission,* which investigated human rights violations under apartheid, and introduced housing, education, and economic development initiatives designed to improve the living standards of the country. In 1996 he oversaw the enactment of a new democratic constitution. While in office Mandela also earned a reputation as an international peacemaker, helping to mediate

North Africa

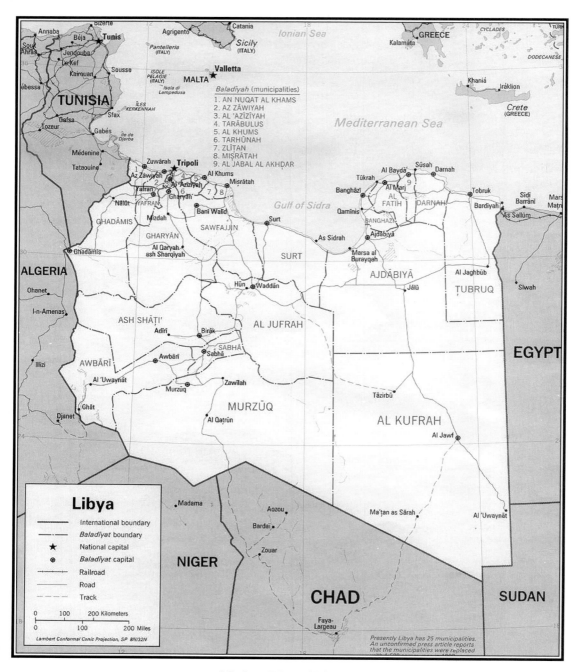

Modern Map of Libya

both in Africa and abroad. That is why his abrupt decision to step down came as such a shock, there was still so much work left to be done. But he was unwavering in his decision that "the time has come for me to take leave."

Although Mandela's sudden departure from the political arena sent shock waves throughout South Africa and the rest of the world, his resignation as his country's first black president was not his only controversial decision that year. A few months earlier, on October 22, 1997, he attended a banquet hosted by Colonel Muammar al-Qaddafi, the leader and head of state of Libya, North Africa's wealthiest country. Not only did he attend the banquet, he delivered a speech especially prepared for the occasion. He began with this greeting:

> I want to thank my dear brother, Colonel Qaddafi, and the government and people of the Great Socialist People's Libyan Arab Jamahiriya for the warm welcome I and my delegation have been given. It is indeed an honor and a privilege to be so received here in Tripoli, and to have the opportunity to renew, in person the strong bonds between our nations.

For many, especially those unfamiliar with the history and politics of Africa, Mandela's speech was a paradox. For some it was viewed as an act of outright betrayal. Here was a man who had built a reputation as a peacemaker, who had become for millions and millions of people around the world a symbol of freedom and justice, now embracing a man regarded as an international symbol of evil. In the years since coming to power in September 1969 Qaddafi had been accused of everything from harboring terrorists to cold-blooded murder. What, if anything, did these two men, with such disparate reputations, have in common?

Judging from Mandela's remarks, the two men, along with their respective countries, shared a deep, personal bond:

> Our visit to your country, brief as it has to be, has proved a moving experience. The people of Libya shared the trenches with us in our struggle for freedom. You were in

Black Orchestra, Libya, 1925 *By the twentieth century about 97 percent of Libya's population were Arabic-speaking Muslims of mixed Arab and Berber descent. However, the trans-Saharan slave trade, which continued during the Ottoman Empire, introduced black Africans and their culture into many Libyan tribes, especially in the Fezzan, the southwestern sector of Libya, and in Tripolitania. Their languages are those of the central Sahara and the eastern Sudan; most also speak Arabic and have adopted Islam.*

the front ranks of those whose selfless and practical support helped assure a victory that was as much yours as it is ours.

After several similar statements expressing his heartfelt gratitude for the friendship of the Libyan people, Mandela concluded his speech as follows:

Friends, ladies and gentlemen, on behalf of my country, may I express our gratitude for your fine hospitality. And may I ask you to join me in wishing good health to Colonel Qaddafi and to the people of Libya. May the ties of friendship and solidarity between our peoples flourish in a partnership for peace and prosperity.

Libya

By all accounts, Mandela's comments, like his presence at the banquet, were a genuine gesture of friendship, but it was a gesture that many people in the international community did not understand. This friendship between two of Africa's most prominent leaders, one a Nobel Prize winner and the other one of the West's most vilified enemies, was bewildering. Mandela's visit and the speech he made to honor the occasion, not to mention Qaddafi's reciprocal hospitality, revealed an image of Libya that challenged many of the West's preconceptions about one of the least familiar countries of the Maghrib and its controversial leader. Just who was this man whom the Western world despised but one of the most renowned leaders of the twentieth century embraced and the developing nations of the world regarded with gratitude?

Muammar al-Qaddafi was born in June 1942 in a goatskin tent in an encampment of the Qadda Berber group in the Sirtic Desert, which separates Fezzan and Tripolitania from Cyrenaica and extends several hundred miles inland from the Gulf of Sidra. His actual birthplace is a Libyan town named Surt. His father, Abuminiar al-Qaddafi, and his mother, Aisha, were members of a nomadic ethnic group that followed the growth of grasses to feed the cattle and sheep they raised. Life was difficult in the drifting sands of the desert where Muammar, the youngest of four children, spent his childhood. It was a childhood in which both tradition and religion helped shape his vision of the world.

Religion and tradition, the two guiding forces in his life, were represented by his Sunni Muslim upbringing and the stories he heard about the death of his grandfather, who had been killed fighting the Italian invasion in 1911. The strict Islamic Bedouin way of life profoundly influenced his later asceticism or austere way of life as well as his political philosophy. In an interview he once stated that growing up Bedouin, learning the harsh nomadic way of life, helped him discover "the natural laws, natural relationships, life in its true nature, before life knew oppression, coercion and exploitation." His upbringing and education in strict Muslim schools also gave him a strong sense of being Arab in general and Libyan in particular. His

"Reform Committee," Misratah, Libya, 1919 *In 1911–1912 the disintegrating Ottoman Empire surrendered to Italy its last North African territories—Tripolitania and Cyrenaica (today's Libya). For many Arabs, this was a betrayal of Muslim interests to the infidels. But Italy's annexation of Libya (1912) was meaningless to the Bedouin tribesmen, who continued their war for the next 20 years. For these Bedouins, unencumbered by any sense of nationhood, the struggle against the colonial power was defending Islam as well as their life-style within their tribal lands.*

This photograph shows Suwaythi, a Tripolitanian nationalist (seated left), and Suleiman Baruni, a Berber and former member of the Turkish parliament (seated right), leaders of the "Reform Committee" that attempted to unify opposition to Italy's claim to Libya. The formation of this committee dates the beginning of the modern Libyan nationalist movement.

powerful belief in Arab nationalism has been a driving force since his early days in the desert.

One of his early idols was Gamal Abdel Nasser, former president of Egypt, and a fervent Arab Nationalist. In fact, when Qaddafi was a young man, both Nasser's nationalist struggle in Egypt and the Palestinian struggle for a homeland in Palestine

Libya

drew him to Arab populist politics. Egyptian teachers and Nasser's regular radio broadcast from Cairo, *Voice of the Arabs,* fueled Qaddafi's passion for nationalism when, as a teenager at fourteen, he was expelled from school in Tripoli for organizing a student strike in support of Nasser's fight with Great Britain over the Suez Canal.

Following his graduation from the Libyan military academy, Qaddafi enrolled in Great Britain's Royal Army Signal School near London. Few Arabs studying abroad wore their traditional dress, but to demonstrate his pride in his culture, Qaddafi wore his. He paraded proudly through the streets of London dressed in his flowing robes and Arab headdress. Qaddafi returned to Libya as an army captain.

After the expulsion of the Germans and Italians from Libya following World War II, the country was controlled by the Allied forces, primarily Great Britain, France, and the United States. Eventually, Libya came under the direction of the United Nations pending the country's independence. When Libya did finally achieve its independence on December 24, 1951, Mohammed Idris Al-Sanusi, or King Idris, became the country's leader. Qaddafi served loyally under King Idris for several years, receiving regular promotions until he attained the rank of major.

These personal achievements could not mask for long Qaddafi's growing sense of dissatisfaction with King Idris's reign, partly because the king continued to remain uncommitted to Arab nationalism and partly because of corruption with the Idris regime. In addition, oil had been discovered in the Libyan Desert in 1959, causing Libya to skyrocket from being one of Africa's poorest countries to one of the wealthiest nations in the world, and Qaddafi and his fellow officers did not think that enough of this wealth was being spent on building up Libya's military—the only way the nation could become totally independent from outside control. In 1961 Qaddafi and several of his fellow officers at the Libyan Military Academy in Benghazi founded a military group called the *Free Officers Movement.* Many of these same men would later help overthrow the Libyan monarchy.

On September 1, 1969, at a time when Arab Nationalist sentiment had reached a fevered pitch in Libya, The Free Officers Movement deposed King Idris in a two-hour bloodless coup. King Idris was out of the country at a Turkish spa for medical treatment at the time of the coup. Five years later, he was tried in absentia on charges of corruption and found guilty. He remained in exile in Cairo, Egypt, until his death on May 25, 1983.

Some historians have called Qaddafi the engineer of the coup, others have characterized him as simply a participant. Even though the new government, known as the *Revolutionary Command Council (RCC),* was initially headed by former political prisoner Mohmud Sulayman al-Maghrabi, the young Qaddafi rose quickly through the ranks of the new government's powerful military and at age twenty-seven, he had de facto control of Libya. The extent of power sharing during the early period of Qaddafi's rule, however, remains a subject of debate but Qaddafi and the RCC patterned their revolution after Nasser's coup in Egypt, and, like Nasser, they saw Arab unity as their goal, their driving force. Unity is a quest Qaddafi pursues to this day.

Within weeks of the takeover, Qaddafi sent representatives to Egypt and the Sudan to propose unity for the three states. Soon after, he proposed a merger of Egypt and Syria in the *Federation of Arab Republics,* and another with Tunisia. All of these efforts failed to gain support. At home Qaddafi's nationalism was more successful.

Once in power, Qaddafi immediately began the overhaul of the Libyan government and society. He charged many of the nation's former leaders with treason, and thousands of people, including Jews and Italians, were tried and imprisoned, or expelled from the country. He promoted Muslim asceticism by banning alcohol. Although he denounced communism for its atheism, relations with Western powers worsened. Even with promises of neutrality, the United States and Great Britain abandoned their military bases in Libya. The final pullout from the U.S. Whellus Air Force Base became the occasion of a Libyan national holiday in June 1970.

Libya

In 1973 Qaddafi instituted *People's Committees* to enable citizens to directly control local and regional government. The *General People's Congress* took over as the national representative body from the RCC, and Qaddifi became the general secretariat of the *Socialist People's Libyan Arab Jamahiriya* or "state of the masses," thus becoming the nation's ultimate decision maker and military leader. Several years later he created *Revolutionary Committees* to guide the People's Committees, and took the title "Leader of the Revolution." He also nationalized the oil and banking industries as well as a large proportion of the retail sector.

The new government took the revenue appropriated through the nationalization of oil and banking industries and used it to improve the standards of living in the country. For the first time, the government distributed oil money to take care of people's basic needs. In fact, during the 1970s and 1980s almost the entire working population in Libya was on the government payroll. The flow of oil revenues ensured that the vast bulk of the population lived comfortably. Housing was practically free, and the government built apartments and modern houses for all who needed them. Water and electricity came without charge.

But this great prosperity did not come without cost to the country and its people. Qaddafi gradually eliminated foreign investment, abolished private enterprise, prohibited all political parties, and renamed the months of the year. He also supported revolutionary causes around the world, some of which brought great disfavor to him and the Libyan people. The controversy seemed to fuel rather than derail his plans and schemes. In spite of sanctions by the United States and the United Nations, Qaddafi continued to implement his own brand of democracy, and eventually he renamed the country the *Great Jamahiriya*, which, loosely translated, means "ruled by the masses."

As a result of his one-man rule, Libya gradually disappeared behind the face, rhetoric, and dubious actions of its leader while some 5 million ordinary Libyans remained anonymous, their daily lives a mystery. "The outside world thinks that this country consists of a man and a desert," the Libyan playwright Mohammed al-Allasi is quoted as saying, "but we are much more than that." It was only through the actions of leaders like

Nelson Mandela that the world has been given the opportunity to rediscover the "real" Libya.

The Great Socialist People's Libyan Arab Jamahiriya is a nation comprising the former Italian colonies of Tripolitania, Cyrenaica, and Fezzan. It is bordered on the north by the Mediterranean Sea, on the east by Egypt, on the southeast by the Republic of Sudan, on the south by Chad and Niger, on the west by Algeria, and on the northwest by Tunisia. By far one of the largest countries in Africa, Libya is 1,757,000 square kilometers, or 678,000 square miles. This vast land is three times the size of France and twice the size of the state of Texas, with land to spare. Tripoli is the capital and largest city.

Libya is a large country, but about 90 percent of it is made up of barren, rock-strewn plains and a sea of sand, with two small areas of hills rising about 900 meters (3,000 feet) in the northwest and northeast. In the south the land rises to the Tibesti Massif, along the Chad border. There are no permanent rivers or streams. As a result the government has undertaken a number of major irrigation projects intended to ease the country's water shortage. One of these is the *Great Man-Made River (GMMR),* a vast water pipeline estimated to cost over $30 billion. The first of five planned phases in the construction of the GMMR was completed in 1996. The project will eventually tap aquifers of the Sarir, Sahha, and Al Kufrah oases and transport fresh water to Libyan cities and agricultural areas along the Mediterranean Coast.

The identity of Libya as an African nation was formed long before Nelson Mandela delivered his speech and even long before Qaddafi's tenure as "Leader of the Revolution." Trade routes and political federations had long connected parts of Libya to Egypt, to the western Islamic political entity of the Maghrib, and to sub-Saharan Africa. The concept of Libya is nevertheless a fairly recent invention, arguably dating back only to the Italian colonial period. Today it comprises three distinct regions: Tripolitania, the cosmopolitan Mediterranean center of trade; Cyrenaica, historically linked to Egypt and the home of the powerful Islamic Sanusi sect; and Fezzan, linked by desert trade routes to sub-Saharan Africa. The identities of these regions were shaped by their individual relationships to the

Libya

Grain Merchant, Tripoli, 1925   *Tripoli is the capital of Libya. Situated on the Mediterranean coast, it is the nation's largest city and its chief seaport. From 1911 to 1943, Tripoli was under Italian control. From 1943 until its independence in 1951, it was occupied by the British.*

powers that occupied their land over the course of the country's long and complex history.

Until Libya achieved independence on December 24, 1951, its history was essentially that of various ethnic groups, regions, and cities, and of the empires of which it was a part. Derived from the name by which a single Berber group was known to the ancient Egyptians, the name *Libya* was subsequently applied by the Greeks to most of North Africa and the term *Libyan* to all of its Berber inhabitants. Although ancient in origin, these names were not used to designate the specific territory of modern Libya and its people until the twentieth century, nor was the whole area formed into a coherent, political unit until then. Therefore, despite the long and distinct history of its region, modern Libya must be viewed as a new country still developing national consciousness. The two most significant forces governing the creation of modern Libya have been geography and religion.

Geography was the principal determinate in the separate historical development of Libya's three traditional regions. Cut off from each other by formidable deserts, particularly the great Sahara, each retained its separate identity into the 1960s. *Sahara* means "desert" in Arabic, and it is the largest desert in the world. Filling nearly all of northern Africa, it measures approximately 4,800 kilometers (3,000 miles) from east to west and between 800 and 1,200 miles north to south. It has a total area of some 3,320,000 square miles (more than 8 million square kilometers).

Rosita Forbes was an intrepid traveller whose adventurous journey to Kufrah in the Libyan desert (1920–1921) earned her extensive worldwide publicity. She was the first European woman to reach this remote and forbidding oasis. Her description of the expedition, The Secret of the Sahara: Kufrah *(1921) became a best-seller, as did her autobiography,* Gypsy in the Sun *(1944).*

Kufrah is a remote oasis about 30 miles long and 12 miles wide near the center of the Libyan desert. In 1895 this crossroads of ancient caravan routes became the headquarters of the Senussi [Sanusiyan], a militant Muslim fraternity founded in 1837. The Senussi held themselves rigidly aloof from the modern world as well as from all contact with people whose religion was not based on their reading of the Koran. Their fanaticism became legendary, as did their opposition to strangers. As a missionary movement, the Senussi sought to reform the lives of Bedouins and to convert to Islam the non-Moslem peoples of the Sahara region. This mystical group of ultraconservative men helped Libya gain its independence in 1951. The head of the Senussi became the first king of Libya. Idris I ruled with virtually absolute power until he was overthrown in 1969 by a group of army officers led by Muammar al-Qaddafi.

No European had been allowed within 15 miles of Kufrah after the Senussi, who numbered about 3,000, had seized the oasis in 1895. However, events of World War I (1914–1918), the ensuing breakup of the Ottoman Empire, and the emergence of new Arab nations in the Middle East, caused the head of the Senussi to be receptive to Rosita Forbes's journey. After 18 months of preparation, Mrs. Forbes began her trek across the Libyan desert in December 1920. She had a letter of safe conduct from the Senussi. The Italian government, as well as leading tribal chiefs, had helped in putting together this expedition, which at its peak consisted of 17 persons and 18 camels. Her round trip to Kufrah, about 1,000 miles, took 12 weeks.

LIBYA

Aujila (now Awjidah), Libya, 1920–1921   *Aujila is an oasis in the Libyan desert. It is about 250 miles due south of the northeastern Libyan port city of Benghazi. Rosita Forbes wrote that Aujila, a mud-built oasis town, presented a half-ruined appearance as so many of the walls and roofs had been left unfinished because of the lack of money. However, she described it as "the most picturesque Libyan town I know, because it possesses no less than nine mosques whose roofs are covered with many clay domes, looking most attractive among the spindly palms." Her comments on Aujila continued as follows:*

As [Aujila] was the real beginning of our expedition, I should like to review for a moment the general situation in which we were placed. Up till then we had travelled almost unnoticed as a couple of stranger Bedawin under the protection of the Sheibs, obviously poor since we had neither clothes nor stores. . . . In spite of this, we had been severely interrogated at Bir Rassam [the last oasis before Aujila. It took almost four-days to cross the desert from Bir Rassam to Aujila]. When we arrived at Aujila, we were so hungry that the only thing we could think of was food, but to our dismay we soon discovered that nothing could be bought in the town. The inhabitants produced only enough for their own needs and refused to sell anything to strangers whose presence they strongly resented. We were just preparing to go to sleep supperless when Sheib produced some eggs, and we spent a blissful hour poaching them. . . . Luckily, our long-delayed caravan arrived at last, provided and equipped [and the most difficult part of the trek to Kufrah across the harshest part of the Libyan desert began].

The whole country is inhabited by Zouiyas [Zouaouah]. Unfortunately for us, the Zouiyas were known and feared as a bad tribe . . . a fanatical warrior tribe, bitterly hostile to the advent of strangers, suspicious, cruel, and treacherous. For this reason whe had been warned that anything in the nature of detailed surveying would be impossible. Any article of luggage whose use they did not understand instantly aroused their distrust [and] even of our original retinue. The production of notebook and pencil immediately sealed their lips. One had to ask the name of hill or well with the utmost casualness, and take care not to write it down till it could be done unseen. . . .

In order to judge distance we calculated that our baggage camels (we had no trotting camels) marched 4 kms. or 2 1/2 miles an hour, but we checked this by using passometers and measuring paces. The first time we fixed the passometers on to the camels' legs there was nearly a revolution, but, in the end, by calling them watches, we induced the retinue to interest themselves in the distance they registered. Likewise with the compass. At first the luminous needle, pointing always to the North Star, was looked upon as a dangerous weapon which might bewitch. . . .

Rosita Forbes, 1920–1921 *Rosita Forbes is photographed with her two female slaves, Zeinab and Hauwa. The following is Mrs. Forbes's diary entry for January 1, 1921:*

> We started about 7 a.m., after a breakfast of rice and sardines and two small cups of hot coffee. We ate dates and malted-milk tablets in the middle of the day as we walked, and at night divided a one-man's ration tin of meat and vegetables with one cup of cold sugarless tea, which was very thirst-quenching. With one cup for cooking the morning rice, our water allowance was therefore less than a quart per day. We were obliged to be very careful because, though we had started with fourteen girbas, i.e. one for each native, and three four-gallon fanatis for ourselves, we were losing water daily. The new girbas all leaked. One of the fantis had also leaked, and the soldiers had made this an excuse for drinking the rest while we were asleep. The other two evaporated a third of their contents in the heat. The retinue ate only Asida, a paste made with onions and flour, so they used more water in cooking than we did. All this meant that on January 1, the day we should have arrived at the outskirts of Taiserbo, we had to reduce the daily water allowance to two small cups, barely a pint per person, and there was much grumbling in consquence.
>
>   One of the great difficulties is that one has to carry not only sufficient water for seven days but all the camel feed and firewood as well. When we left Buttafal, four of our eighteen camels were loaded with water, one with firewood, and four with sacks of dates for fodder. Unfortunately the animals were quite unaccustomed to date feeding. One naga nearly died after her first heavy meal and almost every day after we left the well, one or two camels were obviously suffering and had to be relieved of their loads. . . .

LIBYA

Well at Jedabiya, Libyan Desert, 1920–1921   *Rosita Forbes's description of this well follows:*

[We] arrived at the wadi . . . and found large herds of camels grazing round several small encampments of the Magharba tribe, for there is a large well of perennial water in the bed of the wadi. The water is always slightly brackish, and in summer it is very salty. The well is fed from a spring some 15 feet below the surface. The bed of the wadi is barren sand utterly devoid of vegetation. . . .

The Sahara is bordered in the west by the Atlantic Ocean, in the north by the Atlas Mountains and Mediterranean Sea, in the east by the Red Sea, and in the south by a zone of ancient, immobile sand dunes. Second only to the power and influence of Islam, the Sahara has been one of the most significant and imposing presence in the lives of the people of the Maghrib, especially that part of the Maghrib known as Libya.

At the heart of Tripolitania was its metropolis, Tripoli, for centuries a terminal for caravans plying the Saharan trade routes and a port sheltering pirates and slave traders. Tripolitania's cultural ties were with the Maghrib, of which it was a part geographically and culturally and with which it shared a common history. Tripolititanians developed their political consciousness in reaction to foreign domination, and it was from

North Africa

Libya

Tripolitania that the strongest impulses came for the unification of modern Libya.

Ancient Tripolitania existed under Punic and Roman rule. Phoenician traders were active throughout the Mediterranean area before the twelfth century B.C. By the fifth century B.C., Carthage, the greatest of the overseas Phoenician colonies, had extended its hegemony across much of North Africa, where a distinctive civilization, known as Punic, came into being. Punic settlements on the Libyan coast included Oea (Tripoli), Labadah, later Leptis Magna, and Sabratah, in an area that came to be known collectively as *Tripolis,* or "Three Cities." The influence of Punic civilization on North Africa remains deep-seated. The Berbers displayed a remarkable gift for cultural assimilation, readily synthesizing Punic cults with their folk religion. The Punic language was still spoken in towns of Tripolitania and by Berber farmers in the coastal countryside in the late Roman period.

While a Roman province, the fertile region of Tripolitania exported olive oil and traded gold for slaves brought to the Sahara. As in much of North Africa, Roman presence was

The Kaimakaan of Kufrah, 1921 *The Kaimakaan was the second-highest-ranking Senussi official, second only to Emir Idris, their absolute ruler. The Kaimakaan was photographed here with his slave by Rosita Forbes, who spent nine days at Kufrah in the house of the Emir Idris.*

> There were fine carpets, brass and silver work, and the usual masses of clocks, cushions, and European ornaments, and in the large reception rooms one found a row of immense painted chests in which men carry their luggage on long camel journeys.

Mrs. Forbes never publicly revealed the messages she carried to the British and Italian governments from Emir Idris, whom she never met.

> Idris (1890–1983) succeeded his father as head of the Senussi. In 1922 Italy demanded that he disarm his tribal supporters and submit to Italian rule. When he refused, Italian troops captured the Libyan desert oases. Idris fled to Egypt, where he remained until British forces occupied Libya in 1942 during World War II (1939–1945). In 1951 Libya declared its independence. A national assembly proclaimed Idris the king. His government was an oligarchy of wealthy townsmen and powerful tribal leaders, and as Idris I, he ruled with almost absolute control. Younger army officers and members of the growing urban middle class resented Idris's conservative policies. In September 1969, while Idris was at a Turkish spa for medical treatment, Colonel Muammar al-Qaddafi led a successful revolt. Idris remained in exile in Cairo until his death in 1983.

mainly limited to the coast, while the Berber inhabitants maintained autonomous rule inland. A stretch of desert passable only by camel divided Tripolitania from Cyrenaica to the east. By the second century the coast of Tripolitania was linked by trade and Christian culture to western coastal Roman holdings in Tunis.

In the seventh century, Arab armies moved through North Africa, bringing to Arab culture a new religion named Islam, and dreams of conquest. Arab leaders took Tripolitania in 1650 and from there expanded westward into the land called Ifriqiya. Libya and Tunisia were ruled together, first by the Aghlabids and later by the Fatimid dynasty.

In contrast to Tripolitania, Cyrenaica was oriented toward Egypt and the *Mashrig,* or the Eastern Islamic World, as opposed to the Maghrib, or the Western Islamic World. Legend has it that Greeks from the island of Thera founded the city of Cyrene in 631 B.C. It became an intellectual and cultural center, home to schools of philosophy and medicine, and enriched by the fertile hinterlands' production of grain, wine, and the aphrodisiac *silphium,* an extinct medicinal plant. Within 200 years of Cyrene's founding, four more important Greek cities were established in the area: Brace, or Marj; Euphesperides, later known as Berenice, which is now present-day Benghazi; Teuchira, later Arsinoe, which is present-day Tukrah; and Apollonia, later Susah, the port of Cyrene. Together with Cyrene, these cities were known as the *Pentapolis,* "Five Cities."

With the exception of some of its coastal towns, Cyrenaica was left relatively untouched by the political and cultural influences of the powers that claimed it but were unable to assert their authority in the hinterland, or interior. An element of internal unity was brought to the region's various ethnic groups in the nineteenth century by a Muslim religious order, the *Sanusi,* of which King Idris was a representative, and many Cyrenaicans demonstrated a determination to retain their regional autonomy even after Libyan independence and unification.

Fezzan, the third of the distinct territories that forms modern Libya, was less involved with either the Maghrib or the Mashrig. Its nomads traditionally looked for leadership to local

LIBYA

kinship groups that controlled the oases along the desert trade routes. Throughout its history, Fezzan maintained close relations with sub-Saharan Africa as well as with the coast.

Ruled successively by the Umayyads, Fatimids, and a Berber dynasty, Libya was then partially conquered by the Normans in 1146 but soon abandoned to Almohad control. During the following centuries, Libya had various rulers until it was finally conquered by the Ottoman Turkish Empire.

During the fifteenth and sixteenth centuries, Christians and Muslims competed for control over Mediterranean maritime trade. North African ports became increasingly important. In 1510 Spain captured Tripoli and put its naval base under the protection of the *Knights of St. John of Multa.* That same year, the Ottoman leader Khair al-Din seized the port of Algiers. The Barbarossa brothers, as they were called, extended their holdings eastward along the coast until they captured Tripoli in 1551. The Ottomans, seeking to profit from the commerce through Fezzan, sent armies to exact yearly tribute. But Ottoman power remained concentrated mostly along the coast, particularly Tripoli, while the rest of the country that constitutes present-day Libya was controlled by Islamic sects and Berber confederacies.

During the eighteenth century Tripoli was ruled by a series of military leaders, some more independent than others from the Ottomans. One of the more autonomous rulers was Yusuf ibn Ali Karamanli, who extended his control over Fezzan. In the process of building his empire, Karamanli antagonized Great Britain, especially when he assisted Napoleon Bonaparte in his 1799 campaign to conquer Egypt. On the eve of the new century Great Britain and the United States bombarded Tripoli to end the practice established by the Barbarossa brothers of extorting protection in money from merchant and passenger ships.

Over a period of time the region was thrown into civil war. In 1835, feeling its authority threatened, the Ottoman Empire took steps to regain direct control and by 1842, had reestablished its hegemony over all of Libya. But its control would be only temporary and always the object of attack.

In the early nineteenth century in Cyrenaica—a land where wandering mystics known as *marabouts* possessed significant political authority—a Sufi Islamic sect founded by Muhammad bin Ali al-Sanusi (1787–1859) challenged Ottoman authority by taking control of the desert trade routes. Originally founded to promote a purer form of Islam, the Sanusi order built a series of lodges in Cyrenaica that operated as caravan watering holes in addition to serving as monasteries, schools, and social and commercial centers. Under the leadership of Sanusi's son, Muhammad al-Mahdi, the sect's influence increased. For trade caravans traveling between Benhazi and sub-Saharan markets, the Sanusi lodges offered security and provisions. They also created a cohesive social structure. For this reason, these early social foundations are sometimes viewed as the beginning or the prototype of the modern Libyan state. Certainly the Sanusi sect, and not the Ottoman Empire, was what forged a unified resistance to Italian colonization.

In the late nineteen century, European imperialism in North Africa gained momentum, shifting from economic domination to outright conquest. In part because Algeria, Morocco, and Tunisia had been colonized by France, Italy staked its claim to Tripolitania. Great Britain supported Italy, since an Italian-occupied Libya created a buffer state between French territories and British-ruled Egypt. In 1911 Italy instigated a conflict in order to justify its invasion and occupation of Libya. It accused the Ottomans of supplying arms to Arab Bedouins and then, asserting the need to protect its interest, declared war.

In Cyrenaica, Sanusi followers formed a cohesive resistance, launching a military campaign in 1914 that continued after they officially allied themselves with the Ottoman Empire (present-day Turkey) and Germany during World War I. After the war, the Sanusi untimately forced Italy to grant some concessions in Cyrenaica, and Idris I became the Sanusis' recognized political and religious leader.

In 1923 Italy under Mussolini resumed war against the Sanusi, occupying Tripolitania and Fezzan, but encountering strong resistance in Cyrenaica. Using Eritrean soldiers, the Italian army waged brutal warfare in the desert, cutting off supply lines, filling wells with sand, destroying livestock, confining prisoners in concentration camps, and maintaining a patrolled barbed-wire barrier along the Egyptian border. Italy finally occupied Cyrenaica in 1931 after Sanusi leader Umar al-Makhtar was captured and hanged. Cyrenaica and Tripolitania were then joined in a single colony called Libya, leaving Fezzan as a military territory. Italian occupation of Libya lasted until World War II.

Several major battles were fought in the Libyan Desert during World War II (1939–1945). In January 1943 the British took Tripoli, and by the following month they had seized all of Libya. The British occupied Tripolitania and Cyrenaica, while the French held Fezzan. At the conclusion of World War II, the Allied powers debated the future of Italian colonial territories. In the meantime, Idris, who had lived a life of exile in Egypt during the Italian occupation, returned to Cyrenaica, and with British support established an independent state in Cyrenaica. A national assembly, composed of an equal number of delegates from Cyrenaica, Tripolitania, and Fezzan, convened at Tripoli in 1950 and designated Emir Sayid Idris al-Samusi the head of the Cyrenaican government and "king-designate" leader of its Sanusi sect. The assembly published the Libyan Constitution on October 7, 1951. On December 24, 1951, King Idris proclaimed the United Kingdom of Libya an independent and sovereign state. The first elections were held the following February, and parliament met for the first time in March. Libya joined the Arab League in 1953 and the United Nations two years later. In 1963 the constitution was amended to give women the right to vote.

King Idris's reign lasted exactly eighteen years. He had stood in the breach during some of his country's most devastating defeats and glorious triumphs and was recognized as the symbol of Libya's independence. But on the morning of September 1, 1969, a new era in the history of Libya began, when a group of young army officers overthrew the royal government, and, in the words of Muammar al-Qaddafi, "in one terrible moment of fate, the darkness of ages, from the rule of Turks to the tyranny of the Italians and the era of reaction, bribery, intercession, favoritism, treason and treachery, was dispersed. From now on, Libya is deemed a free, sovereign republic under the name Libyan Arab Republic, ascending with God's help to exalted heights."

Street scene, Fez c. 1923

5
Conclusion

In 1945, at the end of World War II, the idea that the countries of North Africa would be independent, self-governing nations within twenty years would have struck most Europeans, and possibly most people of the Maghrib, as inconceivable. Yet, by 1962 the inconceivable and the unimaginable had become the undeniable. And the reality of such a dramatic transformation shook the very foundations of the Maghrib itself and sent shock waves throughout the rest of the continent and the rest of the world, especially the conquering powers of the world. For the people of the Maghrib region, the dawn of independence was one of the most significant events since Phoenician traders had embarked on their shores. At the very least it meant a redefinition of the relationship that had existed for centuries between the conquerors and the conquered—and marginalized—people of the world. It meant that, possibly for the first time, the history of the Maghrib would be written from the perspective of the people who inhabited the land and no longer as a celebration of European military conquest. Instead, it would be a celebration of the rediscovery of the thousands of years of culture, history, religion, and language that, together with geography, the people of the Maghrib share as common heritage.

But this journey to independence did not come without cost. The road to freedom and self-definition was long and winding and perilously steep. It demanded sacrifice,

NORTH AFRICA

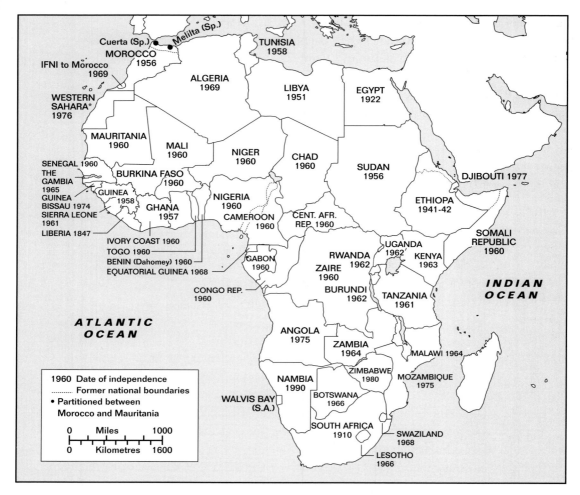

Africa after Independence, 1991

unyielding determination, and commitment in all areas of life, including social, political, economic, religious, and militant resistance from the very first days of European conquest and occupation. And it took diverse paths from diplomatic pleas for reform to civil disobedience through strikes, boycotts, and public demonstrations, to full-scale armed rebellion. Thousands, even

Conclusion

hundreds of thousands of lives were lost in the process. It was a terrible price to pay, but it was the price freedom demanded.

Ironically, once freedom was achieved, the people of North Africa discovered that the battle was not over but just beginning. Economic independence, political stability, and social justice have presented many challenges. Such challenges are inevitable when any people become independent after so many years of, in the words of Kwame Nkrumah, "imperialism, exploitation, and degradation." The people of the Maghrib take heart in the words of Nelson Mandela:

> After climbing a great hill, one only finds that there are many more hills to climb. I have taken a moment here to rest, to steal a view of the glorious vista that surrounds me, to look back on the distance I have come. But I can rest only for a moment, for with freedom comes responsibilities, and I dare not linger, for my long walk is not yet ended.

WORLD WITHOUT END

DEIRDRE SHIELDS

ONE SUMMER'S DAY in 1830, a group of Englishmen met in London and decided to start a learned society to promote "that most important and entertaining branch of knowledge—Geography," and the Royal Geographical Society (RGS) was born.

The society was formed by the Raleigh Travellers' Club, an exclusive dining club, whose members met over exotic meals to swap tales of their travels. Members included Lord Broughton, who had travelled with the poet Byron, and John Barrow, who had worked in the iron foundries of Liverpool before becoming a force in the British Admiralty.

From the start, the Royal Geographical Society led the world in exploration, acting as patron and inspiration for the great expeditions to Africa, the Poles, and the Northwest Passage, that elusive sea connection between the Atlantic and Pacific. In the scramble to map the world, the society embodied the spirit of the age: that English exploration was a form of benign conquest.

The society's gold medal awards for feats of exploration read like a Who's Who of famous explorers, among them David Livingstone, for his 1855 explorations in Africa; the American explorer Robert Peary, for his 1898 discovery of the "northern termination of the Greenland ice;" Captain Robert Scott, the first Englishman to reach the South Pole, in 1912; and on and on.

Today the society's headquarters, housed in a red-brick Victorian lodge in South Kensington, still has the effect of a gentleman's club, with courteous staff, polished wood floors, and fine paintings.

AFTERWORD

The building archives the world's most important collection of private exploration papers, maps, documents, and artefacts. Among the RGS's treasures are the hats Livingstone and Henry Morton Stanley wore at their famous meeting ("Dr. Livingstone, I presume?") at Ujiji in 1871, and the chair the dying Livingstone was carried on during his final days in Zambia. The collection also includes models of expedition ships, paintings, dug-out canoes, polar equipment, and Charles Darwin's pocket sextant.

The library's 500,000 images cover the great moments of exploration. Here is Edmund Hillary's shot of Sherpa Tenzing standing on Everest. Here is Captain Lawrence Oates, who deliberately walked out of his tent in a blizzard to his death because his illness threatened to delay Captain Scott's party. Here, too is the American Museum of Natural History's 1920 expedition across the Gobi Desert in dusty convoy (the first to drive motorised vehicles across a desert).

The day I visited, curator Francis Herbert was trying to find maps for five different groups of adventurers at the same time from the largest private map collection in the world. Among the 900,000 items are maps dating to 1482 and ones showing the geology of the moon and thickness of ice in Antarctica, star atlases, and "secret" topographic maps from the former Soviet Union.

The mountaineer John Hunt pitched a type of base camp in a room at the RGS when he organised the 1953 Everest expedition that put Hillary and Tenzing on top of the world. "The society was my base, and source of my encouragement," said the late Lord Hunt, who noted that the nature of that work is different today from what it was when he was the society's president from 1976 to 1980. "When I was involved, there was still a lot of genuine territorial exploration to be done. Now, virtually every important corner—of the land surface, at any rate—has been discovered, and exploration has become more a matter of detail, filling in the big picture."

The RGS has shifted from filling in blanks on maps to providing a lead for the new kind of exploration, under the banner of geography: "I see exploration not so much as a question of 'what' and 'where' anymore, but 'why' and 'how': How does the earth work, the environment function, and how do we manage our resources sustainably?" says the society's director, Dr. Rita Gardner. "Our role today is to answer such

North Africa

questions at the senior level of scientific research," Gardner continues, "through our big, multidisciplinary expeditions, through the smaller expeditions we support and encourage, and by advancing the subject of geography, advising governments, and encouraging wider public understanding. Geography is the subject of the 21st century because it embraces everything—peoples, cultures, landscapes, environments—and pulls them all together."

The society occupies a unique position in world-class exploration. To be invited to speak at the RGS is still regarded as an accolade, the ultimate seal of approval of Swan, who in 1989 became the first person to walk to both the North and South Poles, and who says, "The hairs still stand on the back of my neck when I think about the first time I spoke at the RGS. It was the greatest honour."

The RGS set Swan on the path of his career as an explorer, assisting him with a 1979 expedition retracing Scott's journey to the South Pole. "I was a Mr. Nobody, trying to raise seven million dollars, and getting nowhere," says Swan. "The RGS didn't tell me I was mad—they gave me access to Scott's private papers. From those, I found fifty sponsors who had supported Scott, and persuaded them to fund me. On the basis of a photograph I found of one of his chaps sitting on a box of 'Shell Spirit,' I got Shell to sponsor the fuel for my ship."

The name "Royal Geographical Society" continues to open doors. Although the society's actual membership—some 12,600 "fellows," as they are called—is small, the organisation offers an incomparable network of people, experience, and expertise. This is seen in the work of the Expeditionary Advisory Centre. The EAC was established in 1980 to provide a focus for would-be explorers. If you want to know how to raise sponsorship, handle snakes safely, or find a mechanic for your trip across the Sahara, the EAC can help. Based in Lord Hunt's old Everest office, the EAC funds some 50 small expeditions a year and offers practical training and advice to hundreds more. Its safety tips range from the pragmatic—"In subzero temperatures, metal spectacle frames can cause frostbite (as can earrings and nose-rings)"—to the unnerving—"Remember: A decapitated snake head can still bite."

The EAC is unique, since it is the only centre in the world that helps small-team, low-budget expeditions, thus keeping the amateur—in the best sense of the word—tradition of exploration alive.

AFTERWORD

"The U.K. still sends out more small expeditions per capita than any other country," says Dr. John Hemming, director of the RGS from 1975 to 1996. During his tenure, Hemming witnessed the growth in exploration-travel. "In the 1960s we'd be dealing with 30 to 40 expeditions a year. By 1997 it was 120, but the quality hadn't gone down—it had gone up. It's a boom time for exploration, and the RGS is right at the heart of it."

While the EAC helps adventure-travellers, it concentrates its funding on scientific field research projects, mostly at the university level. Current projects range from studying the effect of the pet trade on Madagscar's chameleons, to mapping uncharted terrain in the south Ecuadorian cloud forest. Jen Hurst is a typical "graduate" of the EAC. With two fellow Oxford students, she received EAC technical training, support, and a $2,000 grant to do biological surveys in the Kyabobo Range, a new national park in Ghana.

"The RGS's criteria for funding are very strict," says Hurst. "They put you through a real grilling, once you've made your application. They're very tough on safety, and very keen on working alongside people from the host country. The first thing they wanted to be sure of was whether we would involve local students. They're the leaders of good practice in the research field."

When Hurst and her colleagues returned from Ghana in 1994, they presented a case study of their work at an EAC seminar. Their talk prompted a $15,000 award from the BP oil company for them to set up a registered charity, the Kyabobo Conservation Project, to ensure that work in the park continues, and that followup ideas for community-based conservation, social, and education projects are developed. "It's been a great experience, and crucial to the careers we hope to make in environmental work," says Hurst. "And it all started through the RGS."

The RGS is rich in prestige but it is not particularly wealthy in financial terms. Compared to the National Geographic Society in the U.S., the RGS is a pauper. However, bolstered by sponsorship from such companies as British Airways and Discovery Channel Europe, the RGS remains one of Britain's largest organisers of geographical field research overseas.

The ten major projects the society has undertaken over the last 20 or so years have spanned the world, from Pakistan and Oman to Brunei and Australia. The scope is large—hundreds of people are currently working in the field and the emphasis is multidisciplinary, with the aim to break down traditional barriers, not only among the different strands of science but also among nations. This is exploration as The Big Picture, preparing blueprints for governments around the globe to work on. For example, the 1977 Mulu (Sarawak) expedition to Borneo was credited with kick-starting the international concern for tropical rain forests.

The society's three current projects include water and soil erosion studies in Nepal, sustainable land use in Jordan, and a study of the Mascarene Plateau in the western Indian Ocean, to develop ideas on how best to conserve ocean resources in the future.

Projects adhere to a strict code of procedure. "The society works only at the invitation of host governments and in close co-operation with local people," explains Winser. "The findings are published in the host countries first, so they can get the benefit. Ours are long-term projects, looking at processes and trends, adding to the sum of existing knowledge, which is what exploration is about."

Exploration has never been more fashionable in England. More people are travelling adventurously on their own account, and the RGS's increasingly younger membership (the average age has dropped in the last 20 years from over 45 to the early 30s) is exploration-literate and able to make the fine distinctions between adventure / extreme / expedition / scientific travel.

Rebecca Stephens, who in 1993 became the first British woman to summit Everest, says she "pops along on Monday evenings to listen to the lectures." These occasions are sociable, informal affairs, where people find themselves talking to such luminaries as explorer Sir Wilfred Thesiger, who attended Haile Selassie's coronation in Ethiopia in 1930, or David Puttnam, who produced the film *Chariots of Fire* and is a vice president of the RGS. Shortly before his death, Lord Hunt was spotted in deep conversation with the singer George Michael.

Summing up the society's enduring appeal, Shane Winser says, "The Royal Geographical Society is synonymous with exploration, which is seen as something brave and exciting. In a sometimes dull, depressing world, the Royal Geographical Society offers a spirit of adventure people are always attracted to."

CHRONOLOGY

800–2000 B.C.	Western North Africa moves from isolation to connection with the peoples of the Mediterranean and Western Europe. As in those regions, agriculture and pottery are important new technologies in North Africa.
6500 B.C.	The peoples of North Africa live in caves or rock shelters, moving from place to place in search of food. They make stone tools to aid them in the work of hunting and gathering. They also make bone tools and engrave the shells of ostrich eggs to serve as decorated water containers.
5550 B.C.	Impressive images of animals and human hunters are engraved and later painted on rock surfaces. Scholars have not developed a method for determining when this "rock art" was made, but they have outlined categories for classifying it. The large "Wild Fauna" style features hunting scenes with big game, including the giant buffalo. The "Bouidian Pastoral" style refers to images of domestic herds, thought to have been drawn by early farmers.
5000 B.C.	The peoples of Western North Africa become part of trade networks and of the cultural community of the Western Mediterranean.
2500 B.C.	The North African rock art of this period continues to depict animal art but also places a new emphasis on the human figure, equipped with weapons and adornments.
1800 B.C.	The peoples of North Africa build *dolmens* (chambers) of huge stones associated with tombs for the burial of the prominent dead.
1500 B.C.	The climate of the Sahara region undergoes a gradual but drastic change. The encroaching desert separates the peoples of the southern Maghrib from those in the North.
1200 B.C.	Phoenicians establish a colony near the site of present-day Tunis.
814 B.C.	Tradition says this is the year Dido and her companions founded the city of Carthage on the coast of present-day Tunisia.
500 B.C.	Carthage emerges as one of the richest and most powerful cities in the Mediterranean. Numidian kingdoms begin to form beyond Carthaginian frontiers.
264–241 B.C.	The First Punic War is waged.
238 B.C.	Masinissa becomes king of the United Numidia.

CHRONOLOGY

218–202 B.C.	The Second Punic War is waged.
149–146 B.C.	The Third Punic War is waged. Carthage is destroyed.
112–105 B.C.	Jugurtha is defeated by Rome.
46 B.C.	Caesar defeats the followers of Pompey and his ally, King Juba of Numidia. Juba's son is taken to Rome.
29 B.C.	Carthage is refounded.
25 B.C.	Juba II becomes king of Numidia and Mauritania.
1–1500	During this period North Africa is one of the most prosperous and stable regions of the Roman Empire, supplying the capital with staple crops and luxury goods. Peace and wealth create ideal conditions for a flowering of artistic and intellectual life.
40	Ptolemy, son of Juba II, and king of Mauritania, is murdered in Rome.
150	Carthage becomes again a thriving and populous city.
193	Septimius Serverus, a native of Leptis Magna in modern Libya, becomes Roman emperor.
200	Carthage becomes a center of learning.
313	The Donatist schism begins in the African Church.
355	Donatus dies in exile.
395	St. Augustine is named bishop of Hippo.
405	Donatism is officially declared a heresy.
410	Goths capture and loot Rome for three days.
429	Geiseric, king of the Vandals, arrives in Africa.
439	Tunis is invaded by Vandals.
455	Vandals plunder Rome.
533–534	Vandals are defeated.
622	Muhammad and his followers leave Mecca for Medina.
632	Muhammad dies.
644–656	Muslim armies conquer parts of North Africa.
670	The city of Kairouan is founded in Tunisia.
712	Arabs invade Spain.
800	Kairouan becomes a center of religious learning.
808	The city of Fez is founded.

CHRONOLOGY

912 and 947	The first two Fatimid capitals are established in North Africa, al-Mahdiya in 912, and Mansuriya in 947.
969–1050	Fatimid rule of the Maghrib ends. Until 1152 the Zirids and Hammadids, two Berber dynasties, rule Tunisia and eastern Algeria.
1062-1070	The city of Marrakesh is founded.
1062–1150	The Almoravids conquer Morocco.
1082	Morocco is united.
1146	Tripolitania and Cyrenaica are invaded by Normans from Sicily.
1150–1269	Almohad dynasty reigns.
1229–1574	The Hafsids gain control of eastern Maghrib with the decline of Almohads.
1258–1465	The Merinides dynasty reigns.
1453	The Byzantine Empire in Constantinople collapses.
1497	Arab Muslims are driven out of Spain.
1520–1660	The Saadians dynasty reigns.
1535	Tunis is captured by Emperor Charles V.
1573	Tunis becomes a Turkish province.
1574	The Ottoman Empire defeats Spain.
1664–1672	Moulay Ismail reigns.
1711	The Karamanli dynasty reigns.
1797	Tunisia and the United States sign a treaty of peace and friendship.
1801–1805	Barbary Wars are waged.
1817	Piracy is outlawed in Morocco.
1830	France invades Algiers and annexes Algeria.
1871	Algerian rebels are defeated.
1881	The French Protectorate of Tunisia is established.
1906	A European conference is convened to determine the fate of Morocco.
1907	The Young Tunisian Party is formed. France occupies Casablanca.
1911	Spain sends troops to Morocco.

CHRONOLOGY

1912	Tangiers becomes an international zone.
1920–1934	R.F. War is waged against France.
1927	Sidi Muhammad ben Youssef becomes sultan.
1929	Tripolitania and Cyrenaica come under Italian rule.
1934	The Neo-Destour Party is formed.
1938	Habib Bourguiba is imprisoned.
1942	Bourguiba is transferred to France.
1943	Germans are driven out of Tunisia.
1949	Cyrenaica gains independence.
1952	Widespread violence breaks out against French rule.
1951	King Idris declares the independence of the United Kingdom of Libya.
1956	Tunisia gains independence.
1956	Morocco gains independence.
1957	Tunisia becomes a republic. Bourgubia is elected president.
1957	Muhammad V is crowned King of Morocco.
1961	King Muhammad V dies. His son, King Hassan, ascends to the throne.
1962	Morocco adopts its first constitution.
1962	Algeria gains independence.
1963	Ahmed Ben Bella is elected first president of Algeria. The last French people leave Tunisia.
1965	Colonel Houari Boumedienne overthrows Ben Bella.
1969	Colonel Muammar Qaddafi overthrows the royal government of Libya.
1975	An attempted coup against Qaddafi fails.
1982	Temporary headquarters of the PLO are set up in Tunisia.
1982	The United States imposes an embargo on Libya.
1986	The United States bombs Tripoli and Benghazi.
1997	Nelson Mandela visits Libya.
1999	Former foreign minister Abdelaziz Boueflika is elected president of Algeria.
2000	Habib Bourguiba, former president of Tunisia, dies.

FURTHER READING

Africana 2002 Encyclopedia.

Brett, Michael, and Elizabeth Fentress. In Parker Shipton (ed.). *The Berbers*. Maulden: Blackwell Publishers, 1997.

Chamberlain, M. E. *The Scramble for Africa*. New York: Pearson Education, Inc., 1999.

Cockburn, Andrew. "Libya: An Inside Look After 30 Years of Isolation." *National Geographic,* November 2000, pp. 2–31.

Encarta 2002 Encyclopedia.

Encyclopedia Britannica Online.

Gottfried, Ted. *Libya Desert Land in Conflict*. Brookfield: The Millbrook Press, 1994.

Humphries, Rolfe. *The Aeneid of Virgil.* New York: Charles Scribner's Sons, 1951.

Lawson, Don. *Libya and Qaddafi*. New York: Franklin Watts, 1987.

Mansfield, Peter. *The Arabs*. New York: Pelican Books, 1978.

Pakenham, Thomas. *The Scramble for Africa.* New York: Avon Books, 1991.

Palmer, R. R., and Joel Colton. *A History of the Modern World*. New York: Alfred A. Knopf, 1965.

Patai, Raphael. *The Arab Mind*. New York: Hatherleigh Press, 2002.

Rogerson, Barnaby. *A Traveller's History of North Africa.* New York: Interlink Books, 1998.

Sterns, Peter N. (ed.). *The Encyclopedia of World History*. Boston: Houghton Mifflin Company, 2001.

Wepman, Dennis. *Africa: The Struggle for Independence*. New York: Facts On File, Inc., 1993.

RESOURCES USED

Africana 2002 Encyclopedia.

Cockburn, Andrew. "Libya: An Inside Look After 30 Years of Isolation." *National Geographic,* November 2000, pp. 2–31.

Cohen, Mark I., and Lorna Hahn. *Morocco: Old Land, New Nation.* New York: Frederick A. Praeger, Publishers, 1966.

Encarta 2002 Encyclopedia.

Encyclopedia Britannica Online.

Gottfried, Ted. *Libya Desert Land in Conflict.* Brookfield: The Millbrook Press, 1994.

Humphries, Rolfe. *The Aeneid of Virgil.* New York: Charles Scribner's Sons, 1951.

Lawson, Don. *Libya and Qaddafi.* New York: Franklin Watts, 1987.

Mansfield, Peter. *The Arabs.* New York: Penguin Books, 1978.

Nelson, Harold D. (ed.). *Algeria, A Country Study.* Washington D.C.: The American University, 1985.

Nelson, Harold D. (ed.). *Libya, A Country Study.* Washington D.C.: The American University, 1979.

Nelson, Harold D. (ed.). *Morocco, A Country Study.* Washington D.C.: The American University, 1985.

Pakenham, Thomas. *The Scramble for Africa.* New York: Avon Books, 1991.

Palmer, R. R., and Joel Colton. *A History of the Modern World.* New York: Alfred A. Knopf, 1965.

Rogerson, Barnaby. *A Traveller's History of North Africa.* New York: Interlink Books, 1998.

Sterns, Peter N. (ed.). *The Encyclopedia of World History.* Boston: Houghton Mifflin Company, 2001.

INDEX

Across the Sahara from Tripoli to Bornu (Vischner), 47
Aeneas (Dido's love), 70-71
Aeneid (Virgil), 67, 69
Africa, 56, 57, 69
    after independence (1991), *110*
    Carthage, as center of Roman, 75
African National Congress, 87
Africa Nova (New Africa), 77
African Tripolis. *See* Tripoli
Africanus, Publius Cornelius Scipio (Scipio Africanus the Elder), 73
Africa Vetus (Old Africa), 77
Aghlabid Dynasty, 75, 77
Ahab, King, 67
Ahmad Bey, 81
Ahmed I al-marr-sur, 58
Aisha (Muhammad's wife), 52
Algeciras, Act of, 61
Algeria. *See also* Algiers
    Almohad control of, 14
    annexation by France of, 36, 81
    Berbers in, 16. *See also* Berbers
    Biskra (c. 1880), *23*
    caravan (c. 1880), *29*
    colonialism in, 23. *See also* Colonialism
    Constantine (c. 1880), *66*
    crops, 23
    devastation caused by War of Independence in, 26-27
    geography, 21, 27, 28
    guerrilla military offensive throughout, 21, 23
    independence of, 26, 37
    modern map of, *22*
    nomad encampment at oasis of Biskra (c. 1880), *24*
    rock paintings of Tassili N'Ajjer, 28
    Roman ruins in present-day, 31
    War of Independence, 26-27, 37
Algiers. *See also* Algeria
    asks Barbarossa brothers to help defend it, 80. *See also* Piracy
    European trade with, 36
    French occupation of, 23
    as pirate port, 36
    proclaims itself part of Ottoman Empire, 36
    street scene (c. 1923), *20*
    street scene (c. 1890), *37*

Ali (Muhammad's son-in-law), 52, 56
Ali, al-Husayn ibn, 80
Allasi, Mohammed al-, 95-96
Allied powers, 82, 93
All Saints' Day insurrection, 21, 23
    as unyielding struggle against French rule, 26
Almohad Empire, 14, 35, 78
    collapse of, 36
    disintegration of, 57
Almoravid Empire, 14, 35, 75
    expansion of, 57
al-Amin ir. *See* Muhammad
Ammar, Tahar ben, 83
Antony, Mark, 50
Arabian Nights, 20
Arab League, 107
Arabs
    advancement into North Africa by, 33-34, 53, 104
    Berbers form alliance with, 36
    civil war between Berbers and, 57
    conquests of, 33, 53
    destruction of Carthage by, 69. *See also* Carthage
    end of Tunisian supremacy of, 78
    French perception of, 45
    North African invasions by, 14
Arafat, Yasir, 84
Archabas (uncle of Dido), 67
Arianism, 32-33
Arius, 33
Atlas Lands, 16
Atlas Mountains, 44
Augustine, Saint, 32, 71
Augustus, 77
Axis powers, 82

Baal, 71
Bakr, Abu (father of Muhammad's wife), 52
    conquests and missions of, 52-53
Barbarossa brothers, 80, 105. *See also* Din, Aruj; Din, Khair al
Barbary, 15
    horses, 36
Barbary Coast, 36, 59
Barbary States, 15
Barca, Hamicar, 73
Bardo Treaty, 81
    annullment of, 83
Baruni, Suleiman, *92*

Bedouins, 53, 91, 106
    Senussi interaction with, 98
    war against Italy of, 92
Bekkai, Si M'Barek, 39, 41
Belisarius, 32
    defeat of Vandals by, 52
Belus (king of Tyre), 67
Ben Ali, Zine al-Abidine, 84-85
Berbers. *See also* Dido
    absorption of, into Arab society, 14, 53
    al-Kahina ("The Priestess"), 35
    civil war between Arabs and, 57
    confederations of, 49, 57, 105
    defining characteristics of, 34
    derivation of term, 15
    Djurdjura village (c. 1890), *27*
    Donatism and, 31
    early, 28-29
    early references to, 15, 48
    early religion of, 48-49
    form alliance with Arabs, 36
    form Islamic government, 35
    French perception of, 45
    gift for cultural assimilation of, 49, 103
    girls (c. 1890), *14*
    kingdoms of, 30-31
    languages of, 44-45, 46, 48
    in Libya, 97. *See also* Libya
    linguistic basis, as identification of, 44
    in Maghrib, 27-28. *See also* Maghrib
    medieval empire of, 75
    militant resistance to Arab advancement by, 34
    in Morocco, 44, 45. *See also* Morocco
    relationship of Carthage to, 29
    resistance against Roman power by, 32
    social and political organization among, 30
    successive invasions of, 16
    in Tunisia, 16. *See also* Tunisia
    and varied experiences with Islam, 35. *See also* Islam
    woman (c. 1925), *45*
Beys, 80, 81
Bible, 67, 71
Bocchus, (king of Mauritania), 49
Bonaparte, Napoleon, 105

INDEX

Bourguiba, Habib ibn Ali
  death of, 84
  organization of Neo-Destour (New Constitutional Party) by, 82
  positions held by, 83-84
  returns from exile, 82
Bugeaud, General, 37
Byzantine Empire, 34, 53

Caesar, Julius, 49
  forms province of Africa Nova, 77
Caliphs, 35
Camels, 14, *29*, 47
Caravans, 101
  (c. 1880), *29*
  encampments of, 47
Carthage, 29, 103. *See also* Hannibal; Punic Wars
  complete destruction of, 74. *See also* Arabs
  defeat of, 30, 50
  empire controlled by city-state of, 69
  extension of hegemony of, 48
  history of, 71-72
  legend of, 70. *See also* Dido
  resurrection of, 74
  ruins of (1983), *69*
  today, 71
  war between Rome anad, 73-74
Carthaginian Empire, 69, 75
Cato the Elder, 74
Charles V (Hapsburg king-emperor), 80
Charybdis, 72
Chestnut, Charles, 71
Christianity, 31
  Latin, in North Africa, 75
  rapid spread of, 77
Church of Rome, 31
Cleopatra, 50
Colonialism
  in Algeria, 23
  in Morocco, 42
  Sanusi sect resists, 106
  tumultous changes brought about by, 37
Confessions (Saint Augustine), 32
Convention of Marsa, 81
Cyrenaica, 16, 91, 96, 107. *See also* Libya
  orientation of, 104

  surrender to Italy of, 92
  wandering mystics in, 105-106

Dar al Islam (House of Islam), 33. *See also* Islam
Decatur, Stephen, 36
De Gaulle, Charles, 26
Democratic and Popular Republic of Algeria. *See* Algeria
Democratic Constitutional Rally Party, 84-85
Democratic Republic of Vietnam (DRV), 25
Destour (Constitutional Party), 81-82
Deys. *See* Beys
Dido
  early life of, 67
  flees Tyre, 67, 70
  legend of founding of Carthage by, 69, 70
  tragic story of, 70-71
Dien Bien Phu, Battle of, 25
Din, Aruj, 79, 80. *See also* Piracy
Din, Khair al (Redbeard), 79, 80, 105. *See also* Piracy
  mounts seaborne assault against Tunis, 80
Donatism, 31-32
  Numidia, as center of, 77
Donatus, Aelius, 31
Dubois, W.E.B., 71
Duveyrier, Henri, 46

Eighth Crusade, 69
Elissa, Queen. *See* Dido
Emirs, 57
Er Rif, 44
Ethbaal (king of Tyre and Sidon), 67
Exploration of the Sahara (Duveyrier), 46

Fatima (Muhammad's daughter), 52, 56
Fatimid Empire, 57, 77, 104
Faure, Edgar, 83
Federation of Arab Republics, 94
Fez
  as center of learning, 57
  Medersa Bou Inania, main entrance (c. 1890), *38*
  potter's shop (c. 1923), *55*
  street scene (c. 1923), *54*
  washing clothes (c. 1923), *56*

Fez, Treaty of, 62
Fezzan, 90, 91, 96, 104-105, 107
First Indochina War, 25
First Punic War, 72-73. *See also* Punic Wars
FLN. *See* National Liberation Front (Front de Liberation National)
Forbes, Rosita
  description of Aujila, from diary of, 99
  description of house of Emir Idris, 103
  description of well at Jedabiya, 101
  diary entry for January 1, 1921, 100
  round-trip to Kufrah by, 98
  with two female slaves (1920-1921), *100*
France
  Algeria obtains independence from, 27, 37
  annexation of Algeria by, 36, 81
  armed conflicts of, after World War II, 25
  colonial rule in Algeria under, 23
  colonial rule in Morocco under, 39
  concessions granted to Morocco and Tunisia by, 25
  gains part of Morocco, 59, 61
  imperialistic ambitions of, 81
  imposes rule in Morocco, 42
  Moroccan opposition to rule of, 63
  occupies Al Qayrawan, 75
  resistance of Abd al-Qadir to, 36-37
  strained relations between Tunisia and, 84
Frankish kingdom, 77
Free Officers Movement, 93-94
French Indochina, 25

General People's Congress, 95
Geneva Accords, 25
Germany
  involvement in Morocco of, 61, 62
  occupies France and Tunisia, 82
Great Britain
  dealings with Moroccan kaids by, 42
  dispute over Suez Canal between Nasser and, 93

Index

imperialistic ambitions of, 81
involvement in Morocco of, 59, 61
Royal Army Signal School, 93
Great Man-Made River (GMMR), 96
Greeks, 48
   mythology, 72
   Pentapolis ("Five Cities") of, 104
Gypsy in the Sun (Forbes), 98

Hafsid Dynasty, 78
   Sultan Hassan, of, 80
Hajj, 37
Hannibal, 73, 75
   death of, 74
   defeat of, 73
Hanno, Admiral, 48
Haram (king of Tyre), 67
Harems, 20
Hashim confederation, 36
Hassan II, King (of Morocco)
   death of, 64
   Moroccan throne passes to, 63
Hassani, Ismail al-, 58
Hastings, Battle of, 77
Hecateus, 48
Heresy, 32-33
Herodotus, 48, 67
High Atlas Mountains, 42, 45, 61
Homer, 71
Husaynid Dynasty, 80

Iarbas, 70
Idris Al-Sanus, Mohammed. *See* Idris, King (of Libya)
Idris I, 56
Idrisid Dynasty, 56
   end of, 57
Idris II, 56
Idris, King (of Libya), 93, 98, 103, 104
   becomes leader of Sanusis, 106, 107
   death of, 94
   exile of, 103
   reign of, 107
Ifriqiya. *See* Africa
Imazighan, 15-16, 28. *See also* Berbers
Inan, Sultan Abou, 38
Independence
   of Algeria, 26, 37
   of Libya, 93, 97, 103

   of Morocco, 39-41
   road to, for Maghrib region, 109-111
   of Tunisia, 83
   of Western Sahara, 63-64
Independence Party (Hizb al-Istiqlal), 62-63
Indochina, 25
Indochinese Union, 25
Islam, 104
   adoption of, in Libya, 90
   Arabian Peninsula comes under influence of, 33
   Berber and Arab sharing of, 45
   Berber conversion to, 14, 56. *See also* Berbers
   holy cities of, 34, *76*
   influence of, in Morocco, 43
   influence over Berbers of, 49
   motives for expansion of, 53
   spread of, 33-34
   status of women under, 20
   wide variety of Berber experiences with, 35
Israel
   close ties between Tyre and, 67
   establishment of, 58
   Hassan II supports Arab cause in 1967 war against, 63
   migration of Oriental Jews to, 58
Italy
   control of Tripoli by, 97. *See also* Tripoli
   imperialistic ambitions of, 81
   receives North African territories of Ottoman Empire, 92

Jazirat al Maghrib. *See* Maghrib
Jehovah's Witnesses, 33
Jews
   Jewish men (1925), *58*
   Jewish quarter (1925), *60*
   Jewish women (1925), *59*
   migration to Israel of Oriental, 58
   Qaddafi's treatment of Italians and, 94
Jezebel (wife of King Ahab), 67
Juba I, 49
Juba II, 49
   connection to Rome of, 49-50
Jugurtha, 30-31

Kahina, Queen (of Berbers), 35
Kaids, *42*

Karamanli, Yusuf ibn Ali, 105
Kasser Said, Treaty of (Bardo Treaty), 81
Kingdom of Morocco (Al-Maghrib). *See* Morocco
Kinship groups, 105
   Berber, 48
Knights of St. John of Multa, 105-107
Koran, 20, 33
   Islamic college for learning, *38*
   Senussi interpretation of, 98
Krim, Abd el-, 62
Kufrah
   Emir Idris of, 103. *See also* Idris, King (of Libya)
   Kaimakaan of (1921), *102*
   trip of Rosita Forbes to, 98-101

Las Navas de Tolosa, Battle of (Battle of Al-'ugals), 57
Libya. *See also* Tripoli
   during 1970s and 1980s, 95
   Aujila (now Awjidah) (1920-1921), *99*
   beginning of modern nationalist movement in, 92
   Berbers in, 16. *See also* Berbers
   black orchestra (1925), *90*
   bloodless coup in, 94
   British occupation of, 103
   geography, 96
   history of, 105-107
   impact of Sahara Desert on, 16, 101. *See also* Sahara Desert
   independence of, 93, 97, 103
   irrigation projects in, 96
   Italian occupation of, 107
   modern map of, *88*
   nationalization of industries in, 95
   one-man rule by Qaddafi in, 95
   post–World War II control of, 93
   Qaddafi renames, 95
   "Reform Committee," Misratah (1919), *92*
   regions of, 91, 96-97, 103-105
   territories that form modern, 103-105. *See also* Cyrenaica; Fezzan; Tripolitania
   Tuareg chief in southern (c. 1910), *47*
   United Nations sanctions against, 95

Index

Libya. (continued)
    U.S. and British military bases abandoned in, 94
    well at Jedabiya, Libyan Desert (1920-1921), *101*
Libyan Military Academy, 93
Louis IX (king of France), 69

Maghrabi, Mohmud Sulayman al-, 94
Maghrib, 16-17, 43
    Al-Qayrawan chosen as capital of, 34
    early, 27-28
    effect of Sahara Desert on, 101. *See also* Sahara Desert
    geographic differentiation of, 16
    Islamic empire of, 77
    Libya, as one of least familiar countries of, 91
    shared common heritage of, 109
    spread of Islam across, 33
    strategic importance of, 34
Mahdi, Muhammad al-, 106
Makhtar, Umar al-, 107
Mandela, Nelson, 87, 111
    attends and speaks at banquet hosted by Muammar al-Qaddafi, 89-90
    reputation as international peacemaker, 87, 89
Manifesto of Independence, 62
Maquisards, 23
Marabouts, 105-106
Martin-Artajo, 41
Mashriq, 104
Masinissa, 30, 75
    death of, 77
Mecca
    conquest of, 52
    pilgrimage of Abd al-Qadr to, 37
Medina (Medinat-en-Nabi, City of the Prophet), 52
Mendes-France, Pierre, 25, 26
    goes to Tunisia on conciliation mission, 83
Mitterand, François, 24
Mohammed VI, King (of Morocco) (Muhammed ibn al-Hassan), 43, 64
Morocco
    agricultural economy of, 65
    Almohad control of, 14, 57
    architecture, *38*

Berbers in, 16, 44, 45. *See also* Berbers
cities, 38, 41, 42, *54*, *55*, 64-65
conflict between Polisario Front and, 64
as cultural liaison between Islam, Europe, and West Africa, 41, 43
declaration of independence for, 39-41
diversity of, 41
division of, into spheres of influence, 61
early invasions of, 50
Fez. *See* Fez
French concessions granted to, 25. *See also* France
geography, 41, 43-44
golden age of, 58
government of, 43
influx of Moors and Jews from Spain to, 58
invasion by Spain of, 61
modern map of, *40*
nationalism in, 62-63
rebellion in, 62
reunification of, 41
World War II occupation of, 62
Moulay Hassan, Sultan, *42*
Muawiyah I, 35
Muhammad
    death of, 52
    revelations of, 33
    teachings of, 20
Muhammad V, Sultan (of Morocco), 62-63
    assumes title of king, 63
Muslims, 14. *See also* Berbers
    discontent of, under Umayyads, 35. *See also* Umayyad Dynasty
    in Libya, 90
    militant, 98
    Sufi, 37, 106
    Sunni, 91
Mussolini, Benito, 106-107
al-Muwahhid Dynasty, 79

Nasser, Gamal Abdel, 92-93, 94
Nationalism
    beginning of, in Libya, 92
    in Morocco, 62-63
    Qaddafi's passion for, 92-93

Tripolitanian, 92
    in Tunisia, 81
National Liberation Front (Front de Liberation National), 21, 23
    broadcast from Cairo, Egypt, 24
Neo-Destour (New Constitutional Party), 82
    victory for, 83
Neolithic period, 28
New Carthage, 69, 74. *See also* Carthage
New Stone Age, 28
Nicholson, Sir Arthur, 42, 61
Nkrumah, Kwame, 111
Nomads, 14, 91, 105
    encampment of (c. 1880), *18*
    hunters and herders, 28
    at oasis of Biskra (c. 1880), *24*
Norman Conquest, 77
Normans, 77-78
North Africa
    Arab invasions of, 14, 53
    becomes tourist destination, 86
    catastrophic events in, 80-81
    distinctive identity of, 17-19
    elusive definitions of, 15
    geography, 16-18
    influence of geographical circumstances on, 17-18
    influence of Punic civilization on, 28-29, 103
    Islamic dynasties of, 35
    Jews in, 58. *See also* Jews
    Latin Christianity in, 75
    migrating peoples of, 28
    as part of Islamic empire of Maghrib, 77
    postcard picture for tourist shops (1896), *86*
    Roman Empire in, 31, 49-50
Numidia, 30, 75. *See also* Berbers
    Africa Nova (New Africa), 77
    kings of, 49
    Roman interaction with, 31, 49
Nusayr, Musa ibn, 56

Oases, 18, *23*, *24*, 29, 47. *See also* Sahara Desert
    Aujila, *99*
    Saharan well (1896), *79*
    Tunisian, 78
Octavian, 49
Omayyad. *See* Umayyad Caliphate
Oriental Jews, 58. *See also* Jews

Index

Ottoman Empire
  Algiers proclaims itself part of, 36
  end of Algeria as autonomous province of, 37
  gains temporary control of Libya, 105
  Khair al Din appointed regent by, 80
  Morocco escapes influence of, 43
  struggle with Spain over, 80
  surrenders North African territories to Italy, 92
  trans-Saharan slave trade during, 90
  in Tripoli, 105

Palestine
  ancient, 71
  Palestinian struggle for homeland in, 92
Palestine Liberation Organization (PLO), 84
Pétain, Henri Philippe, 62
Phoenicians, 28-29, 67, 71, 103
  trading posts of, 48, 50
  Tyre, 69. *See also* Tyre
Pineau, Christian, 39
Piracy, 36, 79, 80, 101, 105
  Barbary, 59. *See also* Barbarossa brothers
  of Vandals, 50
Polisario Front, 63-64
  conflict between Morocco and, 64
  proclaims Western Sahara independent nation, 63. *See also* Western Sahara
Polybius, 48
Pompey, 49
Ptolemy, 50
Punic civilization. *See also* Phoenicians
  commercial relationships with Berber kinship groups, 48
  influence of, 29, 103
  reaches North Africa, 28-29
Punic Wars, 69, 72, 73-74, 75, 77
Purcell, Thomas, 71
Pygmalion (brother of Dido), 67, 70

Qadda Berber group, 91. *See also* Berbers
Qaddafi, Muammar al-, 98, 103
  accusations against, 89
  becomes leader of Libya, 95
  early life and family of, 91
  education of, 93
  gains de facto control of Libya, 94
  Nelson Mandela attends banquet hosted by, 89
  passion for nationalism of, 92-93
  pursuit of Arab unity by, 94
Qadir, Abd al-, 36
  pilgrimage of, 37
  resistance against France by, 36-37
Qarawiyin University, 57
Al Qayrawan, 34
  Great Mosque (1893), *76*
  architecture of, 75
Qu'ran. *See* Koran

Revolutionary Command Council (RCC), 94
Revolutionary Committees, 95
Rif Republic, 62
Roger II (first king of Sicily), 77
Roman Empire, 30-31
  dissolution of, 33
  Hannibal wages war against, 74
  in North Africa, 31, 49-50
Rome
  Berber conflict with, 30-31
  Berber resistance to, 32
  civil wars of, 30
  domination of Numidia by, 31, 49. *See also* Numidia
  emerging power of, 29
  threat of Donatism to, 31-32
  war between Carthage and, 73-74
Royal Army Signal School, 93
Russell, Charles Taze, 33
Rustamids, 35

Saadians, 57
Sahara Desert, 46, 98. *See also* Oases
  in Algeria, 21
  as Libyan motif, 16
  Moorish cafe (1896), *17*
  nomadic encampment (c. 1880), *18*
  nomadic peoples of, 14, 47. *See also* Tuareg
Sahrawi Arab Democratic Republic (SADR), 63-64
Sallust, 48
Sanusi, Muhammad bin Ali al-, 106

Sanusi sect, 96, 104, 106. *See also* Senussi
  Idris, as leader of, 106, 107
  Mussolini resumes war against, 106-107
Saqigat Sidi Yusuf, French bombing of, 84
Sassanid Empire, 53
Savannas
  in Algeria, 28
  in Morocco, 44
Scipio, Metellus, 49
Scylla, 72
Second Punic War, 73-74, 75, 77. *See also* Punic Wars
Secret of the Sahara: Kufrah (Forbes), 98
Senussi, 103. *See also* Sanusi sect
  fanaticism of, 98
Septimus, palace of (Leptis Magna, 1925), *51*
Severus, Septimius, 77
Shaonia, kaid of (1896), *42*
  fortress of, *61*
Sharifian Empire, 41
  first Sharifian Dynasty, 57-59
Shia, 35. *See also* Islam
Sittius, Publius, 49
Slavery, 31, 101, 103
  abolishment of, in Tunisia, 81
  during Ottoman Empire, 90
  survivors of Carthage sold into, 74
Socialist People's Libyan Arab Jamahiriya, 95
Souks, *55*
South Africa, 87
Spain, 36
  Almohad control of parts of, 14
  Christian reconquest of, 57
  expulsion of Moors and Jews from, 58
  invasion of Morocco by, 61
  Islamic empires conquer portions of southern, 35
  pressure on, to relinquish Spanish Sahara, 63
  receives part of Morocco, 61
  spread of Islam to, 33
  struggle with Ottoman Empire over, 80
Suez Canal
  Nasser's fight with Great Britain over, 93
  opening of, 86

Index

Sufi Muslims, 37, 106
Sufu, 37
Sunni Muslims, 91
Suwaythi, 92

Tahert, 35
Taza Depression, 44
Thapsus, Battle of, 49
Third Punic War, 74. *See also* Punic Wars
Thomas Cook & Son, 86
Tiemcen, 35
Tripoli, 51. *See also* Libya
   as capital of Libya, 96
   grain merchant (1925), *97*
   importance of, 101
   Jewish men (1925), *58*
   Jewish quarter (1925), *60*
   Jewish women (1925), *59*
   power of Ottoman Empire in, 105
Tripolitania, 16, 90, 96, 107. *See also* Libya
   ancient, 103-104
   political consciousness in, 101, 103
   surrender to Italy of, 92
Truth and Reconciliation Commission, 87
Tuareg, 14. *See also* Berbers
   (c. 1880), *46*
   chief, southern Libya (c. 1910), *47*
Tunisia. *See also* Al Qayrawan
   abolishment of slavery in, 81
   Almohad control of, 14
   Arab founding of Al Qayrawan (Kairouan), 34
   Berbers in, 16. *See also* Berbers
   Carthage. *See* Carthage
   derivation of name, 74-75
   early history of modern, 74. *See also* Carthage
   forges alliances with Arab countries, 84
   French concessions granted to, 25. *See also* France
   gains independence, 83
   geography, 78
   German occupation of, during World War II, 82
   modern, 84-85
   modern map of, *68*
   as oasis in the desert, 78
   secularism in, 78
   strained relations between France and, 84
   struggle between Spain and Ottoman Empire over, 80
   Tunis becomes capital of, 79-80
   unrest in, 83
Turner, Joseph, 71
Tyre, 29, 67

Uhud, Battle of, 52
Umayyad Caliphate, 35
Umayyad Dynasty, 35, 56, 57
United Nations, 64
   admission of Tunisia to, 84
   Libya comes under direction of, 93
   Libya joins, 107
   sanctions against Libya, 95
   Tunisians attempt to voice grievances before, 82-83

Vandals, 32, 77
   control of areas by, 50, 52
   defeat of, 52
   destruction by, 51
Vikings (Norsemen), 77
Virgil, 67, 69
   story of Dido, as told by, 70-71
Vischner, Sir Hans, 47
Voice of the Arabs, 93

Walid, Khalid ibn al- ("The Sword of Allah"), 52-53
Western Sahara, 62
   conflict over, 64
   proclaimed independent nation, 63-64
Whellus Air Force Base (Libya), pullout from, 94
William, Duke of Normandy, 77
Women. *See also* Forbes, Rosita
   within Berber society, 45
   receive vote in Libya, 107
   receive vote in Tunisia, 84
   role of, in Donatist movement, 32
   subordinate status of, under Islam, 20
   and succession, in Berber culture, 35
World War II, 62
   armed conflicts of France, after, 25
   battles fought in Libyan Desert, 107
   German occupation of Tunisia during, 82
   occupation of Libya during, 103

Young Tunisians, 81
Youssouf, Abderrahmane, 43

Zama, Battle of, 73
Zirid Dynasty, 75, 77

ABOUT THE AUTHORS

Dr. Richard E. Leakey is a distinguished paleo-anthropologist and conservationist. He is chairman of the Wildlife Clubs of Kenya Association and the Foundation for the Research into the Origins of Man. He presented the BBC-TV series *The Making of Mankind* (1981) and wrote the accompanying book. His other publications include *People of the Lake* (1979) and *One Life* (1984). Richard Leakey, along with his famous parents, Louis and Mary, was named by *Time* magazine as one of the greatest minds of the twentieth century.

John G. Hall received a Bachelor's degree in African American Studies and American Literature from the University of Massachusetts in Boston, and a Master's Degree in Education from Converse College in Spartanburg, South Carolina. He has contributed fiction, nonfiction, and poetry to African Voices, *Aim* magazine, *BackHome, Black Diaspora, Listen* magazine, and *The Sounds of Poetry.* John and his wife Brenda live with their daughter Jessie McConnell-Hall in the mountains of western North Carolina.

Deirdre Shields is the author of many articles dealing with contemporary life in Great Britain. Her essays have appeared in *The Times, The Daily Telegraph, Harpers & Queen,* and *The Field.*